T0184248

SpringerBriefs in Applied Sciences and Technology

PoliMI SpringerBriefs

Series Editors

Barbara Pernici, Politecnico di Milano, Milano, Italy

Stefano Della Torre, Politecnico di Milano, Milano, Italy

Bianca M. Colosimo, Politecnico di Milano, Milano, Italy

Tiziano Faravelli, Politecnico di Milano, Milano, Italy

Roberto Paolucci, Politecnico di Milano, Milano, Italy

Silvia Piardi, Politecnico di Milano, Milano, Italy

Springer, in cooperation with Politecnico di Milano, publishes the PoliMI Springer-Briefs, concise summaries of cutting-edge research and practical applications across a wide spectrum of fields. Featuring compact volumes of 50 to 125 (150 as a maximum) pages, the series covers a range of contents from professional to academic in the following research areas carried out at Politecnico:

- Aerospace Engineering
- Bioengineering
- Electrical Engineering
- Energy and Nuclear Science and Technology
- Environmental and Infrastructure Engineering
- Industrial Chemistry and Chemical Engineering
- Information Technology
- Management, Economics and Industrial Engineering
- Materials Engineering
- Mathematical Models and Methods in Engineering
- Mechanical Engineering
- Structural Seismic and Geotechnical Engineering
- Built Environment and Construction Engineering
- Physics
- Design and Technologies
- Urban Planning, Design, and Policy

Mariacristina Giambruno · Sonia Pistidda

Heritage for a Sustainable Development: The World Heritage Sites and Their Impacts on Cultural Territories

Case Studies from Armenia

POLITECNICO
MILANO 1863

Mariacristina Giambruno (ID)
DASTU - Department of Architecture
and Urban Studies
Politecnico di Milano
Milan, Italy

Sonia Pistidda (ID)
DASTU - Department of Architecture
and Urban Studies
Politecnico di Milano
Milan, Italy

ISSN 2191-530X   ISSN 2191-5318 (electronic)
SpringerBriefs in Applied Sciences and Technology
ISSN 2282-2577   ISSN 2282-2585 (electronic)
PoliMI SpringerBriefs
ISBN 978-3-031-20156-1   ISBN 978-3-031-20157-8 (eBook)
https://doi.org/10.1007/978-3-031-20157-8

This Springer imprint is published by the registered company Springer Nature Switzerland AG
The registered company address is: Gewerbestrasse 11, 6330 Cham, Switzerland

Preface

Armenia has an extraordinary cultural heritage, rich and widespread throughout the territory. However, many threats make its preservation difficult: geological reasons (seismic risk, soil nature, etc.), lack of maintenance, economic and political issues. The rapid growth of tourism occurred in the last decade increased the pressure on heritage, highlighting the contradictions. In fact, from one side, it represents a great opportunity for an economic development and for the communities, a possibility for many territories marginalized for a long time. On the other hand, tourism can put in danger the preservation of sites already tested by long periods of lack of care. It is undeniable that Heritage can have a focus role in the sustainable development but, how is it possible to combine the protection of this invaluable legacy with the needs of development?

The aim of the work is to experiment an integrated approach on the Monasteries of Haghpat, Sanahin, Geghard and moreover, in the Upper Azat Valley to explore possible synergies between different stakeholders involved: local institution, authorities, communities, professionals. The conservation of the Architectural Heritage of these sites is a priority for the identity safeguard but also to start virtuous processes able to ensure a sustainable development for places and communities.

For these reasons, the focus of the research is not only oriented to the so-called "monuments" but also to the minor built heritage that surrounds them: itineraries, landscape, villages can contribute to the construction of tourist paths that start from the main interesting points to go into an immersive experience into the local reality.

The working methodology is organized in some preliminary researches and studies oriented to build a deep and solid knowledge project, able to orient correctly the preservation and design phase. The main goal is to define some working steps able to implement some guiding principles for the correct setting of all the phases.

The book represents the opportunity to systematize the results of a research work commissioned by the World Bank in the framework of Armenia—Local Economy and Infrastructure Development Project.

For each site, a comprehensive master plan has been planned, together with a study of the surroundings to identify strategies to join the preservation of the sites with the creation of possible circuits for the economic improvement of the area. In addition, a broader study was extended to the northern region with the aim to identify itineraries for the promotion of cultural tourism.

Keywords: Built heritage preservation • Knowledge • Integrated approach • Monasteries • Armenia • Haghpat • Sanahin • Geghard • Cultural tourism • Community improvement

Milan, Italy   Mariacristina Giambruno
                                                                                  Sonia Pistidda

Contents

Chapter 1
Cultural Heritage as a Trigger for Sustainable Development in Emerging Countries. The Case of Armenia

Abstract The complexes of Haghpat, Sanahin, and Geghard, three ancient monasteries in Armenia included in the UNESCO World Heritage List, have been an opportunity to reflect on some key issues concerning the role of tourism in the preservation of cultural heritage and the improvement of living conditions of inhabitants. Starting with reflections on the positive and negative effects of these phenomena on the enhancement of the sites, the work presented in the volume aims to provide guidelines for the sustainable development of local communities, starting from preserving their architectural heritage. How can cultural heritage contribute to the sustainable development of sensitive sites? How can we combine the need for transformation with the conservation of local identities? The following chapters explore these complex issues putting cultural heritage at the center of the reflection.

Keywords Cultural heritage and sustainable tourism · World Heritage sites · Relationship between "monuments" and widespread heritage

Cultural heritage represents an undoubted resource for a nation's sustainable development, particularly in emerging countries.

The United Nations 2030 Agenda for Sustainable Development has acknowledged the possible role of cultural heritage, including the need to strengthen its protection and safeguard within the implementation tools of Sustainable Development Goal 11, *Sustainable Cities and Communities.*

Cultural heritage is, in fact, both a public good and a resource, two fundamental factors in ensuring social and economic development.

Firstly, its preservation can reinforce a community's sense of belonging and rooting to a territory and, therefore, its excellent livability. Secondly, cultural heritage can trigger virtuous processes with essential impacts on the population's living conditions. Generally, these benefits are immediately linked to the cultural heritage-tourism binomial and its economic effects. However, these processes may not play any role in the sustainable development of local communities, if the latter are not adequately involved and made aware of their heritage.

On the contrary, tourism can consume a territory and deprive the population of essential resources rather than increase them, if local communities are not protagonists in this process.

However, there are other benefits that heritage conservation can bring to foster the well-being of a population. A preserved "monument" gives quality to a place; it acts as a polarity and can reverberate these positive effects on a village or part of a city (Fiorani et al. 2019; Giambruno and Simonelli 2012; Giambruno and Pistidda 2020; Petrillo and Bellaviti 2018).

The case of Armenia is exemplary in these matters.

The country is rich in architectural and intangible heritage, but for many reasons it has some problems for its conservation: the scarcity of resources; a strong affirmation of the private property and market economy, which were denied for a long time; a vision of restoration that suffers from the country's past isolation.

Tourism development is seen as a possible economic driver and encouraged. However, this process too often excludes the local communities which are not yet ready to welcome visitors according to unavoidable quality standards.

The infrastructure (roads, water, sewage system) is only recently adapting to current standards.

The arrival in the cities of many Russian citizens due to the Ukrainian conflict has changed the housing and consumer goods market, increasing costs. Moreover, the country is still suffering from the conflict with Azerbaijan, with loss of human lives and economic involvement in the war.

Armenia is a two-speed nation: on the one hand there is the capital city, lively and constantly changing, and on the other hand there are the countryside and small towns where inhabitants have been suffering from a lack of work and a shortage of services since the closing of the large Soviet industries.

Therefore, care for cultural heritage and respectful tourism are two possible crucial factors to address the development of this small, young, and courageous nation through the lens of sustainability.

The research presented in this volume starts precisely from these assumptions: to preserve the religious complexes of Haghpat (Figs. 1.1 and 1.2), Sanahin (Figs. 1.3, 1.4 and 1.5), and Geghard (Figs. 1.6 and 1.7) in Armenia, included in the UNESCO World Heritage List, so that they may become the driving force for the sustainable development of the population of the surrounding areas. These communities are not currently taking any advantage of the benefits generated by the tourism activities of the three monasteries. They live in villages, far from the emerging Yerevan, in challenging conditions.

The opportunity for this study was offered by a consultancy assignment awarded by the World Bank in the framework of the *Armenia—Local Economy and Infrastructure Development Project* (P150327).

The aim is to assist the World Bank in designing guidelines for the better implementation of the key interventions identified in the *Bank's Economic Sector Work* (ESW). These objectives include infrastructure improvement; design principles at the site level; an integrated approach in each site, comprising tourism infrastructure

Fig. 1.1 Monastery of Haghpat, general view towards the valley (*Photo* Giambruno 2015)

Fig. 1.2 Monastery of Haghpat, general view of the entrance, 2015

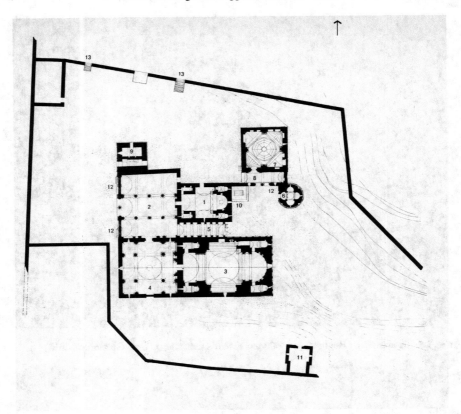

Fig. 1.3 Monastery of Sanahin, general plan of the seventies (1970). (Reproduced from Alpago Novello, Ghalpakhtchian)

blended with site conservation and management; opportunities for increasing visitor spending through an expanded cultural experience; and community inclusion.

The overall objective was to understand the best possible use of a substantial soft loan granted by the World Bank to the Armenian government. Within these general aims, it is significant that the World Bank has thought of reserving a role for cultural heritage, with strong emphasis on tourism, to understand the possibility for Armenia to invest part of said considerable loan in preserving its most significant wealth. The work starts from a solid previous experience in this country, a strong knowledge of architectural heritage problems, and a substantial investigation of the living conditions in the most marginal territories (Casnati 2014).

The main objectives are to provide guidelines for preserving the three monasteries, involve the local population in their protection, and develop the local micro-economies.

This work has been implemented by a team of experts specialized in the conservation and enhancement of monuments, landscapes, and historical urban centers:

Fig. 1.4 Monastery of Sanahin, cross-section of the seventies (1970). (Reproduced from Alpago Novello, Ghalpakhtchian)

an expert in geographic information systems, and a group of structural engineers with a long-run experience in structural repairs, particularly for buildings in seismic areas. The Italian team worked with a group of local professionals: architects, archaeologists, and tourism experts, who guaranteed continuous presence on site.

Therefore, the general objective of this work was to set pilot strategies and guidelines for preserving and promoting cultural heritage as an engine for a compatible and sustainable tourism development that would have a tangible impact on the social and economic development of the country.

The interventions guidelines try to achieve four main objectives:

The study starts from a deep analysis of the opportunities of the three sites and the whole territory, and it suggests solutions to favor better exploitation of the local potential.

It also proposes some ideas to offer various experiences to potential tourists, including less-known aspects of Armenian heritage and culture. These opportunities, if adequately enhanced, can draw a more significant number of visitors and ensure their extended stay in the area and Armenia. The general aim is to guarantee correct heritage preservation, improve the quality of life of the local communities, and promote tourism development.

The Monasteries of Haghpat, Sanahin, and Geghard, must become the center of a virtuous circle. Tourists can enjoy the visit thanks to the necessary facilities and infrastructure. The inhabitants can improve their living conditions by taking

Fig. 1.5 Monastery of Sanahin, general view. (*Photo* Giambruno 2015)

advantage of the opportunities offered by a well-structured international tourism circuit.

Promoting other aspects of Armenian culture and tradition would enrich the tourist experience and create more chances for the involvement of the local population in the earnings from tourism.

The study of the villages surrounding the "monuments", particularly in the cases of Haghpat and Sanahin, is no less critical. These villages not only reflect the population's living conditions but can also offer visitors an appropriate and pleasant environment for their itinerary.

Poor living conditions can force the population to face inadequate situations for the present but also adversely affect the visitors' feelings, conveying them a sense of general insecurity with respect to the place and, therefore, leading them to only choose a short stay there, with minimal earnings for the local community.

In a similar way, improvisation and improper design of the facilities intended to support tourism can cause a general transformation of the historical complex and its surrounding landscape, affecting its preservation and authenticity. The building of incoherent infrastructures could lead to a decrease in the WHS interest and, consequently, to the reduction of attention by cultural tourism agencies which are now very attentive to the care of the places and eager to make travelers enjoy even the minute aspects of local everyday life in tourism destinations.

Fig. 1.6 Monastery of Geghard, general view from east in the seventies. (Reproduced from Sahinian et al. 1973)

The project therefore aims to create employment and income in the rural communities living around the selected WHSs through the sustainable utilization of their cultural and natural assets, the creation of new points of interest, and the realization of the infrastructure required to decrease poverty levels.

During the research phase, the main problems were investigated, both the living conditions of the population (presence of materials detrimental to the health, poor water quality, sewage, and lighting systems), and the cultural heritage assets, both extant and potential.

The continuous exchange with the main stakeholders at the local level, the study of successful examples realized elsewhere, and the assessment of the tourist targets and flows have been crucial for the main goals:

The results of the studies and the project ideas were presented in two public debates to the vice Ministers and delegates of the Armenian Ministries of Culture, Urban Planning, Finance, and others, to collect suggestions and inputs, and were very well received von the whole.

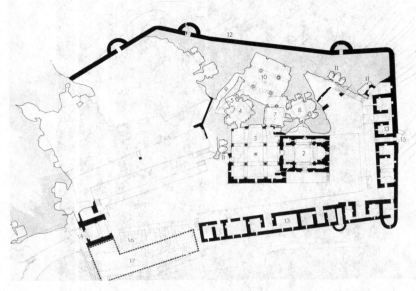

Fig. 1.7 Monastery of Geghard, general plan of the seventies. (Reproduced from Sahinian et al. 1973)

The work ended by handing over to the World Bank and the Armenian government the results of the studies and the guidelines.

The Armenian government, and in particular the Ministry of Culture, was supposed to use it to direct part of the loan received from the World Bank on the main tasks: for conservation works on the three monasteries and the villages, for the creation of tourist infrastructure in their vicinity, for the implementation of services for the population, and for the capacity building of the inhabitants in consciously using the resources of their territory to create small economic activities.

Unfortunately, recent surveys show that the initial objective has not been achieved.

Acknowledgements Although the book is the result of the joint work of all the authors, Mariacristina Giambruno is the particular author of chapters 2 and 4 and Sonia Pistidda of Chaps. 3 and 5. Chapter 1 was shared by both authors.

The group work was composed by Mariacristina Giambruno and Sonia Pistidda with Maurizio Boriani; Gaiané Casnati; Lorenza Petrini; Vincenzo Petrini; Raffaella Simonelli; Roberta Mastropirro; Rosamaria Rombolà; Lucio Speca; Rossana Gabaglio; Francesca Vigotti; Vassilis Mpampatsikos; Nanar Kalantaryan; Lilit Vardanyan; Artur Petrosyan; Siuneh Arakelyan; Gohar Hovakimyan; Kristina Hakobyan.

A special thanks to Gaiané Casnati for allowing the writers to know Armenia. Thanks to her decades of presence in the Country, her studies, and sacrifices to protect its architectural heritage

References

AA VV (1970) Il complesso monastico di Haghbat, Documenti di Architettura armena. Ares, Milano

Alpago Novello A, Ghalpakhtchian O Kh (1970) Il complesso monastico di Sanahin (10–13 sec.), Documenti di Architettura armena. Ares, Milano

Casnati G (ed) (2014) The Politecnico di Milano in Armenia. An Italian Ministry of foreign affairs project for restoration training and support to local institutions for the preservation and conservation of Armenian Heritage. Oemme Edizioni, Venezia

Fiorani D, Franco G, Kealy L, Musso SF, Calvo-Salve MA (ed) (2019) Conservation-consumption. Preserving the tangible and intangibles values. EAAE, Hasselt (Belgio)

Giambruno M, Simonelli R (2012) Cultural heritage preservation and sustainable development. Notes for a shared approach construction. In: Amoêda R, Lira S, Pinheiro C (ed) Heritage 2012. Third international conference on heritage and sustainable development, Porto, Portugal, 19–22 June. Green Lines Institute for Sustainable Development, Barcelos (Portugal)

Giambruno M, Pistidda S (2020) Working for Cultural Heritage. Ethics and preservation in emerging countries from an Italian experience. In: Amoêda R, Lira S, Pinheiro C (ed) Heritage 2020. Seventh International Conference on Heritage and Sustainable Development, Coimbra 8–10 July. Green Lines Institute for Sustainable Development, Barcelos (Portugal)

Petrillo A, Bellaviti P (eds) (2018) Sustainable urban development and globalization. Springer International Publishing, Basel

Sahinian A, Manoukian A, Aslanian AT (1973) G(h)egard. Documenti di Architettura armena. Ares, Milano

Chapter 2
The Relationships Between WHS, the Communities, and the Territories

Abstract Cultural Heritage can play an important and active role in the global challenges such as sustainable development and climate change but it is necessary a change of perspective in its consideration. The enlargement of "heritage meanings", including widespread heritage, the effects produced by the inclusion of a site in the World Heritage List, the role of tourism in a sustainable development of the territories are only some of the issues at the center of the international discussion. The chapter analyzes the state of the art and the consideration of heritage within the future perspectives, try to understand how it can contribute to common goals.

Keywords World Heritage Sites · Communities · Heritage and global challenges · Sustainable development · Climate change

2.1 Widespread Heritage. A Challenge for a Sustainable Future

The meaning of "Cultural Heritage" is currently very complex and cannot therefore be encompassed in a single definition. In the past, the concept was static and closed due to the protection systems implemented by many countries.

Differently, today we need to intend Cultural Heritage as dynamic, ever-changing, and inclusive. Its recognition has to be very rapid, given the speed of our society, and awareness of its uniqueness and irreproducibility of Cultural Heritage is the first step toward its preservation.

Over time, studies and debate by national and international organizations have produced several different definitions. Among these, one of the most convincing and representative is still today that of Cultural Heritage as a "testimony having a value of civilization". This definition was introduced in Italy by the Commissione Franceschini (1967) and internationally by the ICOMOS (1931). These represent essential steps in the evolution of theoretical debate on the matter. For the first time, the general vision of Cultural Heritage was thus extended, shifting from a concept of monument as an "exceptional event" to the broader concept of "widespread" Heritage.

© The Author(s), under exclusive license to Springer Nature Switzerland AG 2023   11
M. Giambruno and S. Pistidda, *Heritage for a Sustainable Development: The World
Heritage Sites and Their Impacts on Cultural Territories*, PoliMI SpringerBriefs,
https://doi.org/10.1007/978-3-031-20157-8_2

This long and arduous has allowed, through an intense cultural debate, to consider the multiple facets of the term "heritage". In this way, many "heritages" have been included in the definition, enabling to extend it infinitely. Indeed, we should consider that every society and every time attribute ever different 'values' to artifacts inherited from the past.

Thanks to this long process, we can now consider many other things as Heritage: evidence of material culture, land uses, disused industrial areas, intangible Heritage (oral expressions, rituals, traditions, crafts, etc.), and urban landscape. All of them are evidence of a silent 'micro-history' without which the memory of great civilizations would not exist.

This minute, silent, and 'discrete' Heritage is fragile for many reasons. Heritage artifacts are often found in marginal areas, made of poor materials, and today it is not easy to imagine a possible use for them, as they require continuous maintenance.

The broadening of the concept of 'heritage' has brought about many issues. Above all, two are crucial: the modalities for its 'recognition' and protection and the strategies for its 'valorization'.

The first involves identifying the many artifacts deserving protection and the implementation of measures for their conservation. The protection imposed by law can only be an effective measure for a limited number of assets. On the contrary, in the case of more extensive and widespread assets, protection needs action to raise awareness of local communities who should take on of the memories of their civilization.

For this reason, we cannot think of Cultural Heritage as something to close in a museum. It must be included in an active and virtuous circuit generating benefits for the places and local communities without jeopardizing their sedimented identity. This is the only way to guarantee continuity in transmitting this Heritage to future generations. This micro-heritage is exposed to several risks on a daily basis: lack of care and maintenance, natural and environmental events, vandalism, tourism pressure, abandonment, and consequent oblivion, but it is also exposed to the hyper-restoration or regeneration, which are too often intended as a replacement.

That Mighty Sculptor, Time by Yourcenar (1983) becomes a double-edged sword for these fragile heritages. On the one hand, it gives them value, adding pages to the 'document'; on the other, it consumes it, subjecting it to the most challenging tests.

The second issue concerns the role of Heritage in contemporary society, and precisely its valorization.

Indeed, 'economic value' is one of the attributes of Heritage, closely related to its uniqueness and unrepeatability. The exact definitions of 'asset' and 'heritage' refer to this.

The issue of 'valorization' of Cultural Heritage highlights many contradictions, and plenty of literature has tried to clarify the difficulties in evaluating these specific assets.

According to some scholars (Settis 2002; Montanari 2014, 2015), valorization exposes Heritage to the most significant risks, because once a potential is identified for attracting financial resources, then conservation needs become secondary.

How to find the correct balance between conservation and valorization?

Tourism and Cultural Heritage are constantly associated with reference to the development processes of territories. But is tourism a driving force for protecting the widespread Heritage, or is it a danger rather than an opportunity?

When economic and natural resources are poor and places are very fragile, the effects of an unbalanced development of tourism in terms of objectives and flows can be devastating. The same is true for apparently more solid contexts that, under uncontrolled tourism pressure, are still at risk of collapsing.

For the weakest local communities, mass tourism is an unexpected disruption. New services arise quickly to meet the increased demand, with no real benefits for the resident population whose living conditions cannot improve in the face of the increase in resource consumption and the significant changes taking place in their territories.

The preservation of Cultural Heritage, which is the true driving force of tourism, becomes secondary while being threatened by the physical consumption caused by the inflow of visitors. On the one hand, "cultural tourism" positively affects places when it reflects visitors' interest in a better knowledge of the sites. On the other hand, the negative effects of this type of tourism on Heritage are unfortunately well known (Della Torre 2015; World Heritage Committee 1998; Harison and Hitchcock 2005).

Among these, the "UNESCO effect" is well known in particular (Gravari-Barbas and Jacquot 2008), which refers to the attraction effect produced by the inclusion of a site on the World Heritage List. The immediate results include risky restoration works to help visitors understand the sites and the expulsion of residents to welcome tourists.

The effects the Airbnb online platform (Gainsforth 2019) has had in the central areas of big cities is a clear example of these processes. As a result of its development, housing rents have increased, leading to the expulsion of the weakest brackets of the local population from the cities' historical centers.

On the contrary, in more marginal places the platform has had beneficial effects, favoring a more settled tourism that has triggered new local micro-economies. These new activities helped maintain a minimum resident population and avoid the abandonment and subsequent degradation of sites.

Based on these assumptions, cultural tourism can be both a resource and a problem. Many examples show how cultural tourism, even when conceived with the best intentions to be 'responsible', 'sustainable', and 'ethical', cannot be the only solution for revitalizing a territory.

Many historic villages undergoing depopulation have associated their social and economic rebirth with tourism and have had a similar destiny. The phenomenon known as 'alberghi diffusi' (scattered hotels) has spread in many territories but with no positive effects on preserving Cultural Heritage and local communities.

The same difficulties are also found in 'experiential tourism', one of the latest travel trends. In this case, the tourist experience is an opportunity to enter into the traditions and culture of a place.

Perfectly aligned to the global culture, the discriminating factor is not the destination but is the type of experience offered. Food and wine tours, cultural itineraries, cooking classes, craft workshops, and sports activities are tailor-made to the places.

In a strange combination of role-playing, the tourist can become a farmer, breeder, artisan, cook, etc.

The many sites that offer this type of package describe it as an 'authentic experience' enriched by the narrative possibility of storytelling and story living. The offer can include routes involving different sectors, such as food and wine, nature, religion, archaeology, and culture. The final goal is to 'surprise' the travelers so that these experiences remains imprinted in their memory over time.

This experience is similar but probably less 'true' compared to the 'responsible tourism trips' which provide complete immersion in the culture and communities of a place.

The desire to give increasingly unique and unrepeatable experiences can erase the places, in a sort of atopy of visitors toward the sites. Does experiential tourism really need a territory and its Cultural Heritage? Or is it only interested in offering more diverse and appealing experiences?

Tourism always has two faces. It is a resource when the promoting policies prioritize the respect of local resources and communities, and when its planning and management involves the local inhabitants with long-term incentive and control policies, with possible correction of the harmful effects of results.

2.2   The Impact of World Heritage Sites Nomination in Emerging Countries

November 2022 will mark the 50th anniversary of the Convention Concerning the Protection of the World Cultural and Natural Heritage. The document is crucial because it brings together Natural and Cultural Heritage protection. These heritages have a decisive role as necessary elements for the development of society and cultural interchange between populations, and thus becomes an indirect instrument of Peace.

More than a thousand properties are inscribed on the World Heritage list as 'Cultural' or 'Natural' sites in 167 countries. Many of these countries are emerging or developing ones, even though the western world and the more affluent nations prevail in the list.

The inscription of a site on the World Heritage List profoundly changes the conservative perspectives and the socio-economic balance of the surrounding territories. Or at least it should be so.

Membership in the World Heritage List should guarantee greater attention by the local governments to protect the sites. Taking care of heritage property is still the responsibility of the individual countries, which are mandatorily required to draw up a Management Plan in the application phase. Moreover, continuative care is needed to prevent the risk of being entered into the "List of World Heritage in Danger" or, even worse, being removed from the World Heritage List.

In addition, it is possible to receive specific consultancy from international experts and to access funding, if even limited, for conservation works, especially for emerging countries.

From the socio-economic point of view, the UNESCO brand is a driving force for the tourism development of a particular area. This aspect is of great interest, and not only for emerging countries to stimulate the economic growth of the marginal regions (Pettenati 2012; Prigent 2013).

Over the last decade, many studies have focused on analyzing this issue, concentrating on the critical aspects of including Cultural Heritage in the World Heritage List, especially in emerging or developing contexts.

Recent studies have highlighted how the conservation of Cultural Heritage is only one (often marginal) of the reasons that drive the different actors (scholars, population, and local authorities) to promote application to the List. Expectations of the different stakeholders are also different: greater effectiveness of protection tools, access to new sources of funding for site conservation, economic benefits in general, and development of the tourism sector in particular (Bagader 2018; Matthys 2018).

From the local communities' perspective, the economic impact of increased tourist flows is the main reason for promoting the nomination, even though the population often pays the highest price with heritage expropriation and exclusion from their territories.

Additionally, the expectation of visitors is certainly higher on World Heritage Sites. The nomination of a site that is little-known outside national borders increases interest from the media, thus bringing it to the international spotlight and increasing global tourist flows.

Tourism pressure involves many critical issues for heritage preservation and local communities.

Indeed, each site has a certain carrying capacity, namely the maximum acceptable number of visitors to avoid irreparable damage to Cultural and Natural Heritage. The improvement of the quality of life of the resident population must to be a priority, and visitors' needs must not conflict with those of the locals.

The "UNESCO effect", as defined by the most recent studies, analyzes the consequences of a "nomination". It is interesting to reflect on it and transfer reflections to the situation of emerging countries.

One first issue regards the conservation of sites. Although there are no detailed studies concerning the effects of inscription in the List on the preservation of sites, observing some of them it is clear that their state of conservation and the interventions carried out are very different and varied. The 1972 Convention rightly provides that the property remains under the responsibility of the State. This leads to economic problems in carrying out restoration works as well as the training of technicians and the workforce who are often not duly up-to-date with contemporary conservation techniques.

In addition, in some cases, the spectacularization of the site prevails in the realization of interventions, for example when a missing part is reconstructed "as it was, where it was" for the specific aim to attract more visitors in this way.

As mentioned above, many studies have analyzed the dynamics and effects of tourism impacts.

The attractiveness of the UNESCO brand appears highly effective in the first years after the nomination but declines in the following years.

The local communities are only occasionally involved in the decision-making processes and tourism development strategies, although the international documents firmly push in this direction. In this way, local communities suffer from a tourism that is only theoretically conceived as 'cultural' and 'sustainable' and too often turns into mass tourism.

The presence of intense tourist flows inevitably affects the local communities and their territory, especially in emerging countries where the resources are just enough to meet the needs of the local inhabitants.

By way of example, think of drinking water, electricity, and health care. In these situations, inhabitants do not benefit at all or only benefit marginally from the money flow generated by tourism. On the contrary, they must compete with visitors in the use their territory's resources.

The *Sustainable Tourism for Development* study by the World Tourism Organization (UNWTO) summarizes very clearly the risks of tourism in developing countries:

- Tourism is a significant and growing contributor to climate change, currently accounting for around 5% of global CO_2 emissions, mainly generated by transport but also by the operation of tourism facilities such as accommodation.
- Local pollution of land and water from poor treatment of solid and liquid waste by tourism businesses and from the activities of tourists can be a problem in some areas.
- Accommodation businesses often use non-renewable and precious resources like land, energy, and water. In some areas, a resort may consume many times more water per person than the local community with which it competes for supply.
- Poorly sited tourism development and inappropriate activities can be very damaging to biodiversity in sensitive areas. Negative impacts to cultural heritage sites can occur where there is poor visitor management.
- Tourism can have negative impacts on local society, through restricting access to land and resources and leading to an increase in crime, sexual exploitation, and threats to social and cultural traditions and values.
- While tourism is well placed to generate accessible jobs, poor working conditions are sometimes found in the sector.
- The economic performance of the sector is susceptible to influences on source markets, such as economic conditions, natural events, and security concerns, although recovery may be rapid when circumstances change.

The Cultural Heritage, inhabitants, and the territory are exposed to obvious risks if the inclusion of a site in the WHL is solely seen as an opportunity to enter into a profitable brand. It is necessary to reflect and return to the spirit of the 1972 Convention, i.e., to protect endangered Heritage by giving it universal interest. This interest means recognizing a country's culture, traditions, and communities.

In Armenia and in the three Monasteries investigated in this study, the described phenomena are minor and with less impact than in other internationally famous sites where economic, political, and social conditions are more critical.

The restoration works are sporadic and not sufficiently suited to the real conservation needs of the Monasteries.

Including the three complexes in the World Heritage List has not brought benefit to interventions' quality and continuity. Moreover, the Management Plan has long been underway and it is still unavailable.

In the specific case of Armenia, there is a difficulty in coordination between the property owners, the powerful Armenian Apostolic Church, and the Ministry of Culture which should supervise and approve the restoration works.

The resources allocated by the property owners are not sufficient for the works needed to preserve the buildings.

This is more visible in the remote complexes of Haghpat and Sanahin than in Geghard, where large parts have recently been restored, unfortunately rebuilding them "in style".

The UNESCO brand has not significantly increased tourist flows. These remained limited and not continuous over time, perhaps also due to the country's weather conditions with freezing and snowy winters. This aspect makes it hard to visit Armenia in winter when the roads—which need upgrades and increased safety—are not easily viable.

As UNESCO's 2014 Periodic Reporting correctly reports about the involvement of local communities, much still needs to be done. They are scarcely involved in decision-making and management processes, and only benefit to a minimum extent from tourism (UNESCO 2014).

In conclusion, in the tiny Republic of Armenia, the "UNESCO effect" is only a potential one and has not been very successful. However, it is necessary to monitor and address it in the best possible way in order to prevent its disruptive effects and bring about the benefits that underpin the recognition of the universal 'value' of Armenian Heritage.

2.3   Heritage in the Global Challenges. Specific Role and Contributions

The recent debate on crucial topics for a sustainable future has been enriched with some essential steps.

The Sustainable Development Goals fostered by the UN Agenda 2030 and the actions to combat Climate Change are only two of the primary measures implemented in recent years.

However, the importance of Heritage in these processes is still poorly recognized and only appears in some marginal passages.

What can the contribution of Cultural Heritage be to addressing these global challenges? The power of Cultural Heritage is universally recognized: as an element of sound identity, it can promote social cohesion, helping to strengthen Peace and well-being. Attracting tourism can empower economic growth and the communities' living conditions, thus contributing to reducing poverty. Moreover, enhancing historic centers can improve infrastructure, guaranteeing better sanitation and access to essential services. The protection of traditional knowledge and skills can help face climate challenges. Direct understanding of Heritage achieved through an immersive experience can support a more inclusive and quality education. Properly enhancing existing buildings is cheaper than building from scratch, and can also increase energy efficiency. Therefore, even if in an indirect way, Cultural Heritage can play a crucial role in the global challenges.

The 2030 Agenda, signed in September 2015 by the governments of 193 UN member countries, identifies 17 Sustainable Development Goals (SDGs) (UN 2015). Explicit reference to Cultural Heritage is only made in Goal 11, "Sustainable cities and communities", with point 11.4 being "protect and safeguard the world's cultural and natural heritage".

The #Culture2030Goal campaign (https://culture2030goal.net/), promoted by some global cultural networks, stresses the insertion of culture in the sustainable development goals but also highlights the scarce attention paid to this issue in worldwide debate.

A recent report by ICOMOS proposes some Policy Guidance to improve the mainstreaming of Heritage in the SDGs. Following the "5 Ps" of the 2030 Agenda (People, Planet, Prosperity, Peace, and Partnership), the document identifies the potential and specific contribution of Heritage in the broader spectrum of actions:

- The knowledge and resources transmitted through Heritage to achieve the well-being of People (SDGs 1, 2, 3, 4, 5, 6, 11);
- a "Culture-Nature" approach and landscape-based solutions to achieve the well-being of the Planet (SDGs 6, 7, 11, 13, 14, 15);
- The shared resources embodied in Heritage to achieve Prosperity of communities (SDGs 5, 8, 9, 11, 12, 14);
- the connecting power of Heritage for social cohesion and dialogue to achieve Peace within and among societies (SDGs 10, 11, 16);
- and the shared medium of Heritage and its connections with all aspects of human life to create Partnerships (SDGs 11, 17) (Labadi et al. 2021).

The slogan adopted by ICOMOS, "Heritage: Driver and Enabler of Sustainability", deserves wider recognition.

Also the UNESCO has recently created a living platform to collect innovative strategies and practices to link heritage conservation with sustainable development (UNESCO 2022).

Climate change represents a challenge for the preservation of architectural heritage. Cultural Heritage and climate change are increasingly at the center of the debate on sustainable development and the circular economy. Cultural Heritage represents a strong identity element, and the use and reuse of existing built Heritage as already available resources is the most apparent aspect of the fight against climate change. But this is not the only issue.

A document prepared by ICOMOS in 2019 proposes to "engaging cultural heritage in climate actions" (ICOMOS 2019a, b, c).

What is the impact of climate change on Cultural Heritage? How can Cultural Heritage contribute to fighting climate change by working on risk management and identifying mitigation strategies?

The threats that can make the natural and cultural environment vulnerable can have many different characteristics: increasing urbanization with consequent abandonment of rural areas, mass tourism, destructions caused by wars, vandalism actions, lack of care, etc. Climate change can multiply and emphasize these risks. The migrations induced by the climate have made communities lose their places and identity and consequently abandon their Heritage. Agricultural production can suffer from climate change and this endangers the livelihood of populations. Changes in climate parameters can also speed up degradation processes of buildings (air pollution, increased temperature, humidity).

Cultural Heritage represents is part of both the problem (it is something to protect vigorously) and part of the solution since it can contribute to increasing the resilience of the communities and places.

This potential is still unexpressed, and the scientific community has not yet profoundly investigated the relationships between Heritage and climate change.

Moreover, as established by the 1972 World Heritage Convention, World Heritage Sites can "act as laboratories of ideas with the potential to set international standards in heritage management".

The ICOMOS Resolution adopted after the meeting in New Delhi in 2017 encourages to "reverse the increase in the global average temperature to well below 2 °C; that adaptation efforts should take into consideration vulnerable communities and ecosystems, and enhance understanding and action with respect to loss and damage from climate change". Knowledge and education can be key factors. Deep knowledge of existing materials and construction techniques can help to evaluate possible behaviors in case of natural risks, simulating models for their mitigation. The same command of knowledge can help identify adaptive strategies to improve the efficiency of buildings. Education means engaging communities, visitors, and stakeholders in taking care of Heritage and fostering proper solutions to the new challenges, reacting to or anticipating the climate risks.

Mitigation methods must not conflict with the preservation of Heritage and traditional practices. For example, energy efficiency of historic buildings can be obtained without jeopardizing their protection.

The issues linked to Cultural Heritage constantly change and evolve rapidly. To bring it into the global debate, we must necessarily consider Heritage as a concept in constant evolution.

Climate change will take us beyond the delicate balance between conservation and development to fundamental questions of human rights and the role of culture in facilitating difficult social Transitions. Heritage practitioners, scholars, educators, and civil society have a central role and urgent responsibility to support communities in safeguarding and advocating for the important roles of cultural heritage in climate action. The multiple and interconnecting layers of climate change Impacts must become a baseline competency of heritage management, as are sustainable development principles (ICOMOS 2019a, b, c).

Heritage management against climate change should become a new discipline. Which elements already contained in cultural Heritage can support climate action? The first step should be identifying the high potential of Built Heritage for prevention and care as continuative practices, which somehow is already a form of mitigation. The rule of the five Rs (Refuse, Reduce, Reuse, Repurpose, Recycle) is hugely relevant for Cultural Heritage. Refusing the superfluous means reusing the great mass of abandoned and dormant Heritage; it means reducing unuseful new constructions and recycling what already exists with a new focus on quality.

The definition of minimum qualitative principles was at the center of the "Cherishing Heritage" conference, organized in the framework of the 2018 celebrations of Cultural Heritage (ICOMOS 2019a, b, c). The document, drafted by a group of experts, tries to encode some necessary everyday actions. One of the critical points regards the need to introduce the "quality" issue in every project step. A final checklist identifies seven main criteria to follow for developing a good project:

1. Knowledge-based. Conduct research and surveys first of all.
2. Public benefit. Keep in mind your responsibility towards society.
3. Compatibility. Keep the 'spirit of place'.
4. Proportionality. Do as much as necessary, but as little as possible.
5. Discernment. Call upon skills and experience.
6. Sustainability. Make it last.
7. Good Governance. The process is part of the success (ICOMOS 2019a, b, c).

Reference to Built Heritage is also poor in the European Green Deal, the action plan implemented by the European Commission for sustainable development. The word heritage appears only in terms of "built and recover in an energy-efficient manner", within the reflection that shows construction as a highly consuming sector (European Commission 2019).

Europa Nostra, with ICOMOS and the Climate Heritage Network (Potts 2021), highlights the importance of considering Heritage an indispensable ally in the challenge against climate change.

Learning from the constructive tradition to repair rather than demolish and rebuild, is more economically sustainable and is the way to connect past, present, and future.

What is the situation in Armenia?

In a voluntary national review report for 2020, the Armenian Government provided some preliminary suggestions for SDGs priorities (the Republic of Armenia 2020). In the description of the main SDGs 2030 for the country, unfortunately, Heritage has no specific recognition, and the focus is more on law reforms, energy diversification, climate change mitigation, and infrastructure. Moreover, regarding Goal 11, Sustainable cities and communities, the report does not mention Cultural Heritage as an intervention element but only points out problems such as reducing inequalities between cities and rural settlements, improving public spaces, and protecting the natural environment.

Armenia is highly vulnerable to climate change, with agriculture, human health, water resources, forestry, and transport among the most affected sectors. According to a World Bank analysis, "it ranks as the fourth most vulnerable country in Eastern Europe and Central Asia regarding climate-change risk". Air pollution, changes in the Lake Sevan ecosystem, and deforestation are only the most evident problems. Armenia ratified the Paris Climate Agreement in 2017, and in 2021 the Minister of Environment and the Food and Agriculture Organization of the United Nations (FAO), signed an agreement "to increase forest cover by 2.5%, reduce the fuel-wood demand of rural communities by at least 30%, enable sustainable and climate-adaptive forest management, and ensure technology transfer to rural communities, the private sector and institutions as part of a forest-energy nexus approach" (United Nations 2021).

A joint report by the World Bank and the Asian Development Bank in 2021 (Climate Risk Country Profile: Armenia 2021) points out the actual scenario and the projections, highlighting the natural hazards. Armenia has significant exposure to risks, including mudflow and landslides, flood risks, reduction of water resources, and drought, especially in rural areas. In the past, earthquakes produced important disasters, especially in 1988. This data clearly shows a worrying scenario if interventions will not be well planned, and urgently. Cultural Heritage can play a crucial role in the future if the Government, the international organizations supporting the country, and all the stakeholders involved in the choices will realize the importance of our legacy from the past.

2.4   The Role of Local Communities and Stakeholders in the Preservation of Architectural Heritage

More than forty years ago, the Amsterdam Declaration (1975) expressly defined, for the first time, the role of local communities in the conservation and transmission of Cultural Heritage, stressing the vision of Integrated Conservation: "Integrated conservation involves the responsibility of local authorities and call for citizen's participation". Cultural Heritage needs recognition and care to survive, and the role of education and culture is crucial: "only if it is appreciated by the public and in particular by the younger generation. Educational programs for all ages should, therefore, give increased attention to this subject" (Council of Europe 1975).

More recently, the Faro Convention (2005) has stressed the need to engage communities as prior guardians of Cultural Heritage, giving them an active and participatory role. "Recognising the need to put people and human values at the center of an enlarged and cross-disciplinary concept of cultural Heritage. (…) Cultural Heritage is a group of resources inherited from the past that people identify, independently of ownership, as a reflection and expression of their constantly evolving values, beliefs, knowledge, and traditions. It includes all aspects of the environment resulting from the interaction between people and places through time" (Council of Europe 2005, Art. 2a).

The Florence Declaration, signed in 2014, has drawn attention on a "Bottom-up approach for effective conservation and management of heritage". The lens of tourism influences the document which, however, points out the role of inhabitants in the care process. "A community with highly developed cultural awareness and the capacity to identify unique cultural values within their community is in a position to be empowered to protect the integrity, authenticity, and continuity of the cultural heritage recognized within that community" (ICOMOS 2014).

Communities have become increasingly important in international documents, and their role is even more remarkable, considering the widespread Heritage and how people deal with it on a daily basis. Culture of care implemented by the Heritage's first users, i.e., local inhabitants, is the main guarantee for the survival of fragile Heritage.

Enhancing the awareness of the local people about their Heritage is the first step to fostering virtuous processes. Recognition is the first obstacle in conservation practices, especially in emerging countries where the cultural memory linked to evidence from the past is not so obvious. Moreover, the profound modifications related to use can increase the threats. Again, education and culture appear to be the only solutions: training people to see Heritage as a resource, working on-site with the communities to raise their awareness and love for their legacy, and spread good practices for using it as a potential source of economic income.

The role of the population in the direct preservation of Cultural Heritage needs rethinking, with active involvement in the decision-making processes.

High poverty rates prevail in Armenia, especially in rural areas. This situation puts in danger the widespread Heritage, firstly because of the modifications made in self-construction works. Splitting of building, addition of new parts, and random maintenance works aimed at adapting the spaces to new uses and needs, are frequent on the historical buildings and are usually carried out by the local inhabitants, with few resources and expertise. This self-construction phenomenon leads buildings to lose their identity and authenticity.

A people-centered approach to Heritage is required to face the new global challenges, as pointed out in the resolution approved by ICOMOS at the end of 2020:

Promote people-centered approaches, the connections of people with Heritage and places; intercultural dialogue and understanding, sustainability and well-being when addressing local, national, and international heritage policies and practice. This will better realize the full potential of cultural Heritage to deliver climate-resilient pathways to strengthen sustainable development while promoting a just transition to low-carbon futures.

Honor heritage communities and individuals' rights, taking into consideration past and future generations, acknowledge their role in developing and implementing resilience strategies in a rapidly changing world, and assist authorities to empower citizens to maintain and develop their values and livelihoods in a dignified, responsible, and sustainable manner.

Work to sustainably synergize cultural heritage conservation and management with the diverse cultural, environmental, and socio-economic concerns of people and communities through building partnerships with relevant sectors, such as social and health services; cultural and creative industries; nature and biodiversity conservation; tourism; urban and territorial planning and development; and infrastructure and energy provision (ICOMOS 2019a).

Communities should connect to Heritage and vice versa. Communities can co-manage the sites and guarantee long-term conservation thanks to their continuous presence in the places. A previous report by ICCROM (2015) identified different types of people involved, underlining that "Communities can be communities of place (those who live within or near to Heritage), communities of interest (those who feel a connection to or are interested in Heritage) or communities of practice (those who work with Heritage)" (ICCROM 2015).

It is urgent to look at these different users as active "change makers". Not considering the community's contribution to Heritage can negatively affect both issues. On the contrary, social engagement can improve social dignity, cultural inclusion, and poverty reduction. The periodic report by UNESCO of 2014 on the World Heritage Sites of Haghpat and Sanahin only partially recognize the role of the local communities pointing that "Local communities have some input into discussions relating to management but no direct role in management" (Art. 4.3.8) and "Indigenous peoples directly contribute to some decisions relating to management but their involvement could be improved" (Art. 4.3.9) (UNESCO 2014).

References

Bagader M (2018) The impacts of UNESCO's built heritage conservation policies (2010–2020) on historic Jeddah built environment. In: WIT transactions on the built environment, vol 177

Commissione Franceschini (1967) Per la salvezza dei Beni culturali in Italia. Atti e documenti della Commissione d'indagine per la tutela e la valorizzazione del patrimonio storico, archeologico, artistico e del paesaggio, Casa Editrice Colombo, Roma. https://whc.unesco.org/en/hul/. Accessed 15 June 2022

Council of Europe (1975) Declaration of Amsterdam, Congress on the European Architectural Heritage. https://www.icomos.org/en/charters-and-texts/179-articles-en-francais/ressources/charters-and-standards/170-european-charter-of-the-architectural-heritage. Accessed 10 July 2022

Council of Europe (2005) Framework convention on the value of cultural heritage for society, Faro, 27.10.2005. https://www.coe.int/en/web/culture-and-heritage/faro-convention. Accessed 10 July 2022

Culture2030goal. Towards a culture goal. http://culture2030goal.net/. Accessed 21 July 2022

Della Torre S (2015) Lezioni imparate sul campo dei distretti culturali. Il Capitale Culturale, Supplementi 03:61–73

European Commission (2019) The European Green Deal. https://eur-lex.europa.eu/legal-content/EN/ALL/?uri=CELEX:52019DC0640. Accessed 18 June 2022

Gainsforth S (2019) Airbnb città merce. Storie di resistenza alla gentrificazione digitale. DeriveApprodi, Roma

Gravari-Barbas M, Jacquot S (2008) Impacts socio-économiques de l'inscription d'un site sur la liste du patrimoine mondial: une revue de la litterature. In: Prud'Homme R, Les impacts socio-économiques de l'inscription d'un site sur la liste du patrimoine mondial: trois études. UNESCO, Paris

Harrison D, Hitchcock M (eds) (2005) The politics of world heritage: negotiating tourism and conservation. Channel View Publications, Cleveden

ICCROM (2015) People-centred approaches to the conservation of cultural heritage. Living Heritage. https://www.iccrom.org/sites/default/files/PCA_Annexe-2.pdf. Accessed 24 July 2022

ICOMOS (1931) The Athens Charter for the restoration of historic monuments. Adopted at the First international congress of architects and technicians of historic monuments. https://www.ico mos.org/en/167-the-athens-charter-for-the-restoration-of-historic-monuments. Accessed 14 July 2022

ICOMOS (2014) The Florence Declaration on heritage and landscape as human values. https://www.icomos.org/images/DOCUMENTS/Secretariat/2015/GA_2014_results/GA2014_Sympos ium_FlorenceDeclaration_EN_final_20150318.pdf. Accessed 3 June 2022

ICOMOS (2019a) Climate change and cultural heritage working group. The future of our pasts: engaging cultural heritage in climate actions, July 1. ICOMOS, Paris

ICOMOS (2019b) European quality principles for EU-funded interventions with potential impact upon cultural Heritage. ICOMOS, Paris. http://openarchive.icomos.org/2083/1/European_Qua lity_Principles_2019b_EN.PDF. Accessed 12 June 2022

ICOMOS (2019c) People-centred approaches to cultural heritage, resolution 20GA/19. ICOMOS, Paris. https://www.icomos.org/images/DOCUMENTS/Secretariat/2021/OCDIRBA/Resolu tion_20GA19_Peolple_Centred_Approaches_to_Cultural_Heritage.pdf. Accessed 24 July 2022

Labadi S, Giliberto F, Rosetti I, Shetabi L, Yildirim E (2021) Heritage and the sustainable development goals: policy guidance for heritage and development actors. ICOMOS, Paris

Matthys A (2018) L'effet UNESCO sur le développement local, Métropolitiques, 17 septembre. https://www.metropolitiques.eu/L-effet-UNESCO-sur-le-developpement-local.html. Accessed 15 July 2022

Montanari T (2014) Istruzioni per l'uso del futuro. Il patrimonio culturale e la democrazia che verrà. Minimum Fax, Roma

Montanari T (2015) Privati del patrimonio. Einaudi, Torino

Pettenati G (2012) Uno sguardo geografico sulla World Heritage List: la territorializzazione della candidatura, Annali del Turismo, 1. Geopress edizioni

Potts A (2021) Executive summary. In: European cultural heritage green paper executive summary. Europa Nostra, The Hague & Brussels

Prigent L (2013) L'inscription au patrimoine mondial de l'UNESCO, les promesses d'un label? Revue Internationale Et Stratégique 2(90):127–135

Republic of Armenia (2020) Sustainable development goals. Voluntary national review report. https://sustainabledevelopment.un.org/content/documents/26318Armenia_VNRFINAL. pdf. Accessed 18 July 2022

Settis S (2002) Italia S.p.A. L'assalto al patrimonio culturale. Einaudi, Torino

The World Bank Group and the Asian Development Bank (2021) Climate risk country profile: Armenia. https://climateknowledgeportal.worldbank.org/sites/default/files/2021-06/15765-WB_Armenia%20Country%20Profile-WEB_0.pdf. Accessed 15 June 2022

World Heritage Committee (1998) Reports on the state of conservation of properties inscribed on the world heritage list. United Nations Educational, Scientific and Cultural Organization convention concerning the protection of the world cultural and natural heritage, twenty-second session, Kyoto, Japan, November 30–December 5

Yourcenar M (1983) Il tempo, grande scultore. Einaudi, Torino

UNESCO (2014) Monasteries of Haghpat and Sanahin, periodic reporting Cycle 2, Section II. file:///C:/Users/sonia/Downloads/PR-C2-S2-777.pdf. Accessed 12 June 2022

UNESCO (2022) World heritage canopy. Heritage solutions for sustainable future. https://whc.une sco.org/en/canopy/. Accessed 12 July 2022

United Nations (2015) Transforming our world. The 2030 agenda for sustainable development. https://sustainabledevelopment.un.org/content/documents/21252030%20Agenda%20for%20S ustainable%20Development%20web.pdf. Accessed 21 July 2022

United Nations (2021) More trees in Armenia to help counter climate change. https://armenia.un. org/en/131545-more-trees-armenia-help-counter-climate-change. Accessed 10 July 2022

Chapter 3
A Journey Through Armenian Heritage

Abstract This chapter describes the whole scenario for developing the proposals for the World Heritage Sites (WHS) Master Plans. The methodological process started with a quick survey to understand the characters and the main problems of the Armenian Architectural Heritage and to insert the work on the monasteries of Haghpat, Sanahin, and Geghard into a more general framework. A second step concerns the tourism analysis of opportunities and threats for an emerging country like Armenia. Some valuable data allow the comprehension of the country's actual situation to plan possible future scenarios. Finally, the working methodology underlying the proposed Master Plans is outlined, starting from some theoretical considerations and highlighting the two main goals: on the one hand, the proper conservation of the three monasteries, and, on the other, the necessary improvement to the living conditions of the resident population, with a general look to the broader context.

Keywords Architectural and urban heritage in Armenia · Preservation policies · Opportunities and threats of tourism · The methodological approach of master plan for WHS of Haghpat · Sanahin · and Geghard

3.1 The Armenian Architectural and Urban Heritage. Beyond the Monasteries

The first impression traveling through Armenia is the omnipresence of Heritage: a large density of historical Architectural Heritage against a contrasting background of low population density.

With over 24,000 cultural and historical monuments, three state reserves (Khosrov, Shikahogh, Erebuni), and two National Parks (Sevan and Dilijan), the country alternates natural landscapes (high and majestic mountains, desert valleys, barren plateaus, lakes, and impressive basalt formations such as those near Gyumri) with significant architectural environments, like monasteries, historical centers, castles, and ancient villages (Ministry of Economy of the Republic of Armenia 2020).

Undoubtedly the best-known historical architectures in Armenia are the numerous monastic complexes spread in the territory that attract numerous tourists from many parts of the world every year (Cuneo 1988).

In addition to the three World Heritage sites explicitly covered by this study, there are other important sites of great interest: Noravank (Fig. 3.1) (Alpago Novello and Leni 1985), Khor Virap (Fig. 3.2), Aruch (AA VV 1986) Akhtala, Tatev, Marmashen, and Kobayr, together with some fascinating churches, such as the Ererouk basilica (Alpago Novello and Paboudjian 1977; Casnati and Tonghini 2012) and the Tegher church, to mention a few.

The notoriety of these sites, also confirmed by the WHL, which in addition to Haghpat, Sanahin, and Geghard, also includes Echmiatsin (the Holy See of the Armenian Church), has not saved them from the two main problems that plague Armenian Architectural Heritage.

Fig. 3.1 Monastery of Noravank, general view of the church (*Photo* M. Giambruno 2012)

Fig. 3.2 Monastery of Khor Virap, general view (*Photo* M. Giambruno 2012)

On the one hand, reconstructions in style and random restorations are the most diffused interventions; on the other, abandonment and ruin condemn many of these monuments to oblivion.

The seismicity of the Armenian territory, and the strong earthquakes that have affected it over the centuries, certainly play an essential role in these two extremes in the interventions on historical buildings. However, we will mention later other accompanying problems.

In addition to the well-known monasteries, Armenia has a varied and stratified architectural heritage covering a period from prehistory to Soviet modernism. The latter is no less important but certainly more endangered in terms of conservation (Harouthiounian 1975; Harutyunyan 2017; Marouti 2018; Casnati 2014).

We do not intend to provide a complete list, which would require an accurate census campaign that the body in charge of protection in Armenia has not yet carried out. However, a brief overview, based on numerous inspections conducted not organically over some years, can be helpful to give an idea of the richness of this territory and to understand the problems affecting local Cultural Heritage.

Areni cave complex (Fig. 3.3), Zorats Karer (megalithic structures dating to the Middle Bronze Age, also known as the "Armenian Stonehenge"), Ughtasar (rock graffiti probably dating between the 5th and 2nd millennium BC), Uytse, Godedzor, and Goris are primary and exciting prehistoric sites.

Fig. 3.3 Areni cave complex, detail of the excavation campaign (*Photo* M. Giambruno 2012)

The Areni cave is a site of great interest, dating back to the Copper Age. The area was recently excavated using international archaeological methods. It consists of three caves where traces of agricultural and cult activities are evident. The large jars still contain the remains of pomegranate seed processing and wine production, as well as some extraordinarily well-preserved burials. The exceptional nature of the site lends itself well to tourism development, although large flows of people could compromise its conservation.

As Boriani rightly suggested (Boriani and Premoli 2012), this place could be part of a network of prehistoric Armenian sites which should be included in a UNESCO tentative list. The need for a management plan and the evaluation of the consequent load capacity could positively affect the conservation of these places.

The Hellenistic period also has its "stone representation" in the Armenian territory. The Garni site was built in the first century AD and includes a temple and the remains of baths. The latter belonged to the palace that once stood in the area. After the destruction caused by an earthquake in the seventeenth century, in the 1960s restoration works were carried out that completely reconstructed the temple from the base to the top (Fig. 3.4). However, the reconstruction is quite recognizable to a trained eye.

The Garni site welcomes many visitors each year thanks to its proximity to Yerevan and its being unique in the panorama of Armenian Cultural Heritage.

Fig. 3.4 The temple in Garni (*Photo* F. Vigotti 2022)

Near Garni an exceptional natural monument is found, the Garni Gorge (Figs. 3.5 and 3.6).

It consists of high hexagonal basalt columns that outline the banks of the valley formed by the river. The site was accessible until a few years ago through a basalt paved path, and is now connected by a new driveway built with World Bank funds. However, better accessibility could damage the site's conservation, which could be reached by many more visitors, with the inevitable problems that this could entail (parking, garbage, and damage to the basalt columns). In contrast to the large tourist flow, the village of Garni stands out for its lack of quality, centrality, and services. Just think that its streets are still only partially paved.

Leaping in time to demonstrate the stratification of the Armenian Architectural Heritage and briefly outline its main problems, we can get to the centuries of the Silk Road. The caravanserai of the Selim Pass (Fig. 3.7) is one of the most significant testimonies among the best-preserved caravanserai system. Near the building, now accessible by a driveway, the remains are found of the ancient path and some of the bridges that connected the route. The caravanserai urgently needs conservation works, especially in the interior spaces, but above all, it requires a reuse that is coherent and compatible with its characteristics. These works could prevent constant vandalism.

Fig. 3.5 The natural monument of Garni Gorge, general view (*Photo* S. Pistidda 2022)

Fig. 3.6 The natural monument of Garni Gorge, detail of basaltic columns (*Photo* M. Giambruno 2022)

Fig. 3.7 The caravanserai of the Selim Pass (*Photo* M. Giambruno 2012)

Approaching contemporary times, the Architectural Heritage of the Soviet-era deserves a special mention.

The presence of the "communist regime" is widespread in the territory, so much so that even the smallest villages have a building dating back to this period. Today, several factors threaten this Heritage.

Although the well-known phenomenon of the *damnatio memoriae* is not so strong in Armenia, on the one hand, accepting such a recent heritage as a cultural asset is difficult. The other threat is the loss of the original use of many of these buildings due to the country's rapid modernization, especially in the capital.

In Yerevan, for example, there are metro stations, such as those in Republic Square (Fig. 3.8) or Isahakyan Street (Fig. 3.9), as well as many government buildings and funicular stations, such as the one on Koryun Street (Fig. 3.10), that are now in a state of abandonment and ruin (Harutyunyan 2018).

In the suburbs, many condominiums dating back to socialist modernism are exciting examples of that historical period and of the first experiments of building unification and prefabrication (Fig. 3.11). In these cases, the inhabitants, to improve their living conditions in houses that, being experimental, had several defects, have been profoundly changing the original architecture with small but repeated transformations.

The story of the Rossiya cinema is a further testimony of what is happening to this Heritage.

Fig. 3.8   Yerevan, the metro station in Republic Square (*Photo* S. Pistidda 2022)

Fig. 3.9   Yerevan, the metro station in Isahakyan Street (*Photo* M. Giambruno 2014)

The cinema, built between 1968 and 1975, could seat 2500 people. The building stood out for its large reinforced concrete roof. Abandoned after Armenia's independence from the Soviet Union, it subsequently became a shopping center and all traces of the previous interior spaces were completely erased.

Fig. 3.10 Yerevan, the funicular station in Koryun Street, now completely abandoned (*Photo* S. Pistidda 2022)

Fig. 3.11 Yerevan suburbs, detail of a building of a modernist district (*Photo* M. Giambruno 2014)

Fig. 3.12 The Sevan Writers' Resort on the Sevan Lake (*Photo* S. Pistidda 2022)

The fund granted by the Getty Foundation (Keeping It Modern fund) for studying the Sevan Writers' Resort (Fig. 3.12) represents a further sign of the need to develop strategies for preserving modernist Heritage. The complex on Sevan Lake is undoubtedly an exciting example of the formalization of a socialist utopia in architecture and is now in a state of neglect.

However, those known as historical centers are the testimonies of the past that seem to suffer more in Armenia, between abandonment and demolition. Armenia is, in fact, rich in cities that still retain their oldest parts intact.

To mention a few examples: Goris, with stone houses featuring wooden balconies; Dilijan, although turned into a little Disneyland by a recent intervention on Sharambeyan Street (Fig. 3.13) sponsored by the Tufenkian Heritage Foundation; and also Ashtarak, a small town north of Yerevan, whose oldest part has been wholly abandoned (Figs. 3.14 and 3.15).

The cities of Yerevan and Gyumri are two exemplary case studies that help better understand the condition of "historical centers" in Armenia.

In the capital, the historic center can be said to have almost wholly disappeared. The low buildings that characterized the historic center were demolished to be replaced by buildings of fifteen and more floors. In most cases, the façade has been preserved as a mere decoration for the new reinforced concrete structures.

Fig. 3.13 Sharambeyan Street in Dilijan (*Photo* M. Giambruno 2015)

Fig. 3.14 Some abandoned buildings in the old town of Ashtarak (*Photo* M. Giambruno 2017)

Fig. 3.15 Some abandoned buildings in Ashtarak, the hamam (*Photo* M. Giambruno 2017)

The situation is entirely different in Gyumri, the country's second-largest city. Designed by the Russians in the nineteenth century, it suffered from a violent earthquake in 1988 that led to its gradual depopulation. The city is still struggling to recover, but many buildings are abandoned, and the urban voids caused by the earthquake are still clearly visible (Figs. 3.16 and 3.17).

However, the Urban Development Committee of the Republic of Armenia (UDC) has recently undertaken studies and research to address updating the Master Plan to preserve the historic city.

Despite that described above, one should not think that the Republic of Armenia has no protection policies. They are based on the Soviet Heritage, which also recognized "minor heritage" as a legacy to protect. For example, the historical part of Gyumri has been entirely included in a Reserve Museum to preserve its integrity and control interventions.

The Constitution of the Young Republic, issued in 1995, assigns the State the duty to protect its historical Heritage, and a specific protection law was enacted in 1998.

Implementing this law is assigned to the Ministry of Culture and its officials, who are in charge of approving and supervising all interventions on architectural Heritage, just like in Italy; however, in the Armenian context problems are different and much more complex than in our Country (Petrosyan and Bădescu 2016).

Firstly, in Armenia the work of officials is complicated by the difficulty of accessing the cadastral maps of the various buildings and the registers of their owners to notify the protection rules. Secondly, there is no specific training for architects in the restoration field. The National University of Architecture and Construction

Fig. 3.16 The airport of Gyumri that recalls the Zvarnots of Yerevan in the architectural forms (*Photo* M. Giambruno 2022)

Fig. 3.17 Gyumri, one of the main street in the historic center (*Photo* S. Pistidda 2022)

of Armenia (NUACA) has only recently considered the opportunity of creating a particular degree program in this sector.

Another problem is that there are no guidelines, tender specifications, or price lists for restoration works. The approval of projects is entrusted to a committee of experts within the Ministry of Culture. Based on what we saw in the meetings we attended, they do not evaluate works according to the compatibility and correctness of the techniques proposed to preserve the material, but evaluate them based on compliance of reconstruction projects to philological criteria.

Finally, the rapid modernization experienced by the country, as in the case of Yerevan, is attracting large capital and leading the entire population to consider the legacy of the past as obsolete and to replace it with new buildings as soon as there is an economic opportunity to do so.

These are phenomena from which no country, in its emerging phase, has remained untouched.

3.2   Tourism as a Driving Force for the Country: Opportunities and Threats

Today, tourism is a driving force in developing countries as it is their primary income source. Cultural and Natural Heritage are among the main attractive elements for visiting a country.

Cultural Heritage and Tourism are often combined topics when discussing the possible development of a territory. The economic benefits generated by the tourism industry are huge, and they can express concrete support for the local communities: employment, stimulation of investment, and opportunities for the disadvantaged areas. Moreover, the positive effects of cultural exchanges and increasing knowledge can bring additional benefits to host communities, also stimulating social and cultural impacts (community engagement, infrastructure development, strengthening of culture and traditions, and social awareness). It can indirectly contribute to the development of environmental awareness and political stability. On the other hand, uncontrolled tourism development can also entail numerous risks for countries with a more fragile and vulnerable Cultural Heritage. Large flows of visitors can have strongly stressful impact on the natural and cultural resources of places, generating a rapid development of new settlements to meet the increasing tourism demand.

Looking at different contexts that have undergone these phenomena, it is clear that the real benefits of these processes for the resident population have been modest compared to the corresponding depletion of resources.

In 1995, the first World Conference on Sustainable Tourism held in Lanzarote, Canary Islands, stressed these risks (WTO World Tourism Organization 1995).

Almost three decades later, many of the concepts expressed have not yet been implemented, and there is still thought of increasing tourist flows.

The effects of tourism in emerging countries can be devastating: replacement of traditional productive activities, alteration of the environment, consumption of natural resources such as water and energy, and the rising cost of living.

These phenomena are unfortunately often linked to the inclusion of a site into the World Heritage List.

The ICOMOS International Cultural Tourism Committee (ICTC), a specialist International Scientific Committee of ICOMOS, has long been involved in these issues.

The first Charter of Cultural Tourism (ICOMOS 1976) recognized the threats of this irreversible phenomenon, stressing the priority of preserving Cultural and Natural Heritage.

The extent of the growing phenomenon required the setting of "Principles and Guidelines for Managing Tourism at Places of Cultural and Heritage Significance", contained in the International Cultural Tourism Charter of 2002 (ICOMOS 2002).

The main goal of the document is to provide principles to manage the complex relationship between the host community and the visitors, based on a cooperative approach for the protection of the cultural and natural environment. One of the relevant aspects is the extension of the reflection from monuments to "all forms of cultural heritage", using the broad concept of "Heritage significance" and the need to communicate it.

Six are the main principles for the guiding actions: 1. Encouraging public awareness of Heritage; 2. Managing the dynamic relationship; 3. Ensuring a worthwhile visitor experience; 4. Involving host and indigenous communities; 5. Providing benefits for the local community; 6. Responsibility of promotion programs.

Some evaluation questionnaires help to verify whether the principles are satisfied. The revision process initiated by the Florence Declaration led to a new document in 2021 with important extensions and specifications of the recommended principles (ICOMOS 2021), introducing tools to increase the resilience of communities and Cultural Heritage and integrating climate action and sustainability measures in the management. The World Tourism Organization (UNWTO) has long been working on the concept of sustainable tourism, recognizing that community engagement is the main factor for a shift in perspective.

During the last decades, a significant number of conferences have produced interesting suggestions, charters, and guidelines that have not always resulted in concrete policy strategies at the local level but, in any case, have been helpful in keeping the debate alive. Alongside the definitions of sustainable tourism, the new concepts of ecotourism and responsible tourism have entered the discussion.

The Ministry of Economy of the Republic of Armenia identifies the role of tourism as a priority in the country's sustainable development, recognizing that only a conscious development can generate long-term positive effects. It entrusts to the Tourism Committee the responsibility of strengthening the sector by collaborating with other stakeholders (government partners, international organizations, universities, etc.).

The outlook for tourism and its contribution to economic growth is quite optimistic, and it has grown significantly in the last ten years. Statistics report a 14.9%

increase in arrivals between 2018 and 2019, with a higher incidence in the summer months. The first 34 countries by arrivals see Russia, Georgia, and Iran in the first places, highlighting a not yet significant number of international tourists. The visitors from Russia, Georgia, and Iran mainly come to visit friends and family or for business reasons, still showing a low number of people that come only for tourist reasons (Atoyan 2016). The most visited destination is still the capital Yerevan (Fig. 3.18) with a few other sites such as Garni, Geghard (Figs. 3.19 and 3.20), Tsakhkadzor, Echmiadzin, Zvartnots, Sardarapat, and Lake Sevan (Fig. 3.21). The Armenian diaspora represents the reason of the majority of tourist arrivals (62% in 2016) with long stays (25 days) and repeated visits to the country (Ministry of Economy 2020).

The Travel and Tourism Competitiveness Report drawn by the World Economic Forum (WEF) in May 2022 ranked Armenia 61st out of 117 countries, improving its position by 18 points in comparison to the previous report of 2019, where it was 79th among 140 countries (Uppink and Soshkin 2022).

Among the European countries, the most significant number of arrivals are from Germany (28,655 in 2018 and 39,690 in 2019) and France (27,651 in 2018 and 32,397 in 2019).

Fig. 3.18 The capital Yerevan, the Cascade (*Photo* S. Pistidda 2022)

Fig. 3.19 Monastery of
Geghard, one of the most
visited sites in Armenia
(*Photo* S. Pistidda 2022)

The 2022 Yearbook of Tourism Statistics dataset by the United Nations World
Tourism Organization (UNWTO) (the Republic of Armenia joined it in 2003) also
reports some interesting data (WTO World Tourism Organization 2022).

In recent years, international tourist arrivals have been growing steadily in
Armenia, well above the world average. In 2016, the number of travelers reached
1,259,657; in 2017 they were 1,494,779; in 2018 1,651,782; and in 2019 1,894,377.

The 2020 figure was affected by the Covid-19 pandemic, with only 375,216
arrivals and a decrease by −80.19% compared to 2019. The analysis of arrivals
by country of residence shows a predominance of people from Europe and Eastern
Europe with an increasing number during the years.

According to the WTO 2022 statistics, the primary purpose of the visit is personal
(holidays, leisure, and recreation, with 1,593 presences in 2019), and only a few
arrivals are due to business and professional reasons (301 in 2019). Package tour
arrivals are a minimum figure (in 2019, only 301 on a total of 1,894). Outbound
tourism is mainly for personal purposes.

Fig. 3.20 Monastery of Geghard, sellers at the entrance of the site (*Photo* M. Giambruno 2015)

Fig. 3.21 Lake Sevan (*Photo* S. Pistidda 2022)

According to macroeconomic indicators related to international tourism, inbound tourism expenditure over GDP was 10.6% in 2017, 10.9% in 2018, and 11.4% in 2019. With increased visitors and spending, the benefits to Armenia's economy are also growing, and are expected to continue to grow in the future.

According to the 2018 World Travel Tourism Council Report, the tourism and travel sector directly contributed 4.4% to the country's GDP (477.7 million dollars in 2017). The industry generated 3.9% of total employment (44,500 jobs in 2017), visitor exports generated 29.2% of total exports, and in 2017 the travel and tourism investment was 4.6% of the total investment (World Travel and Tourism Council WTTC 2018). Armenian cultural and historical assets, medieval religious architecture in particular, are a significant draw card in attracting international tourists. However, it is becoming increasingly apparent that the industry's potential is not being fully realized.

Although several iconic sites are quite well presented, numerous smaller heritage locations are less well-known and need substantial improvement, but have much to offer tourists: archaeological sites, little museums, historical urban centers, modern architecture, etc.

The Natural Heritage and its potential as a driver for tourism development (hiking, trekking, bird watching, cross-country skiing, mountaineering, climbing, etc.) are still under consideration. In addition, the potential for agritourism and wine tasting/culinary tourism has been explored only recently.

Let's look at the above figures while considering that we live in a period of economic downturns. We can guess that in the future, the more considerable contribution to the income from the tourism industry will come more from an increase in the number of visitors than from a rise of their spending, also considering the stop in tourism during the Covid-19 pandemic.

As shown by a survey conducted in the Tumanyan province, where the new B&Bs have been suddenly open with full fruition, the demand for diffused and low-cost hospitality is high and is expected to increase. The type of tourists that is expected to increase in the short term are independent travelers, people visiting their families, and people from emerging markets (Brazil, Russia, India, and China can be considered as the most important tourism source market in the world).

While the economic development of Armenia in the years to come is generating a solvable middle class of nationals traveling in the country for leisure and vacations, local tourists are also expected to rise significantly. Therefore, improving tourist facilities and visitor access to new destinations scattered over the whole territory, and exploring new and diversified leisure offers is of major importance. Meanwhile, widening the offer and promotion of sustainable solutions is also the best means to allow an equal distribution of tourism earnings and increase local employment opportunities by involving local communities and their cultural and economic development.

The 2019 report highlighted the importance of investing in the workforce, in particular in their training and the quality of facilities and services offered (Ministry of Economy of the Republic of Armenia 2020). The report pointed out the current main challenges of the tourism industry: acquiring a position in the global market as a tourism destination, investing in travel promotion, develop policies to favor access and infrastructure, and the need for interventions to mitigate seasonality. It is significant that the protection of Cultural Heritage is identified as one of the essential prerequisites for sustainable tourism development.

A study commissioned by the World Bank to investigate a framework for sustainable tourism development in Armenia gave an insight into Armenians' perception of their Heritage and their willingness to participate actively in its safeguard and enhancement. Not surprisingly, it emerged that Armenian citizens are attached to their cultural Heritage in all the accepted meanings of the word, considering monuments and nature as integral to one another. More surprisingly, they visit architectural monuments more frequently than theatre, cinemas, or museums, but slightly less than natural areas.

Armenians underestimate the importance of their distinctive hospitality tradition, which is also a facet of their Heritage and is of high cultural significance despite its being not so clearly visible. They do not perceive hospitality as a characteristic asset of their culture, nor as a possible attraction for tourists who, on the contrary, have demonstrated to appreciate it. Finally, a significant percentage of the respondents said to be optimistic for their future and to feel that they can personally contribute to improving their Heritage.

Not only are they willing to pay for the preservation of Armenian monuments, but their willingness to pay increases as their income also increases. This trend has been confirmed by the outputs of the interviews undertaken in Sanahin and Haghpat. The population started with a distrustful attitude, but then they showed an increasing interest in being involved in activities aimed at developing their villages and creating new businesses in connection with Cultural Heritage improvements. The willingness of people to pay for Cultural heritage conservation is expected to increase as they become more empowered and wealthier.

Domestic visitors also recognize the importance of protecting Armenian monuments and, after the problematic years in the immediate aftermath of national independence, finally feel they have the possibility to dedicate personal resources to their conservation. This aspect makes us believe that the revitalizing the old traditional lottery dedicated to the safeguard of Armenian monuments would be a successful action. As for the international tourists (actual and potential), a poor state of conservation of monuments would considerably and negatively influence their decision to visit the country, with site congestion still not considered a threat.

Armenians and foreign visitors consider the values of Architectural Heritage to be inseparable from the quality of the surrounding environment. However, none of the stakeholders, including on-site interviewees, experts involved in the focus groups, nor travel agents seem to be concerned about the possible overexploitation of the country's cultural and natural resources. They do not consider that sites with monuments might become congested nor that monument conservation could conflict with economic growth objectives. On the other hand, European agencies confirm that their clients consider avoiding congestion as crucial.

To conclude, this overview suggests that cultural tourism in Armenia is a sector with a strong growth potential not only in terms of international tourist flows but also in terms of further development of a domestic market. The opportunity given by tourism growth should be exploited and promoted as an economic, social, and environmental driver. The evidence of what has been realized in other countries shows that a relatively modest investment in Cultural Heritage can pay substantial dividends.

An improved enhancement of the Armenian cultural identity may stimulate growth and employment, favor social inclusion, and increase the interest in Armenia as a tourism destination. However, it is still a challenge to ensure adequate preservation, maintenance, and enhancement of Armenian Heritage, which is often at risk due to geological and economic factors as well as the lack of maintenance and correct restorative interventions. The pressure for rapid economic development and the need for a substantial increase in revenue related to tourism may put at risk the conservation of monuments and works of art. This report aims to find solutions to widen and diversify the offer for tourism targets, thus limiting the potential damages and favoring the spreading of benefits.

With careful strategic planning and relatively modest public sector investments, tourism is one of the sectors that is best able to deliver on employment at a moment when job creation is a priority.

3.3 From Monuments to Widespread Heritage: A Shift in Perspective

The study's methodological approach

Cultural Heritage is the heart of what it means to be Armenian and could increasingly contribute to GDP and improve people's lives and wellness. This consideration is the starting point of the research work presented in this volume.

However, the main objective is linked to the tourist enhancement of the three monasteries included in the World Heritage, their conservation, the management of visitor flows, and the development of a solid reception network. The basic assumption is that tourism cannot properly develop where the population's living conditions are precarious. No responsible tourist is comfortable visiting places marked by poverty and lacking minimum hygiene standard or primary health care services.

This research work has tried to reverse assumptions: exploiting the three religious complexes can be a trigger for global recovery, with a positive impact on surrounding villages and the local population's life in general. Beyond the monasteries, as mentioned above, there is a widespread historical heritage of which Armenia is rich and which is a fundamental element to preserve, not in the form of a museum but rather as a living, current, livable and inhabitable element.

Cultural Heritage represents a vital resource for communities which should be enjoyed first by those living nearby. To achieve this goal, the population should become the first custodian of its Heritage and take an active part in transmitting it to the future. Their being aware that their Heritage can attract visitors and be a primary driver for developing flourishing tourism industry, can improve the life of the local people.

The Faro Convention clarifies this point, stressing that cultural Heritage is what the population recognizes as such, and that its uses and meaning are fundamental.

Even that of Faro remains a declaration of intent, like any Charter (Council of Europe 2005). By the way, Armenia will not have to join it, and the bottom-up approach now almost fashionable in the Western world, faces some structural difficulties in this country. After all, until the recent past, in Armenia the State took care of every citizen's need and also of Cultural Heritage preservation.

Even today, the population, especially in rural areas, tends to believe that someone else has to take care of their things. This phenomenon is an evident legacy of the regime which hinders the development of autonomous initiatives (entrepreneurship, Cultural Heritage preservation), and makes citizens reluctant to organize themselves in labor cooperatives, which are perceived as a legacy of the past.

For this reason, some efforts should also be focused on capacity building and creating incentives to invest time, money, and human resources to promote private investment in Cultural Heritage.

The investment package should include funding for cultural activities, community awareness, public development, SME incubators, training, etc. These actions can promote high quality in transmitting the place's values. Once the preservation of Heritage guaranteed, awareness will be crucial, more than the works on the buildings.

Based on these assumptions, the research work was developed according to two objectives: on the one hand, the correct preservation of the three monasteries and, on the other, the necessary improvement of the living conditions of the resident population (Figs. 3.22 and 3.23).

In particular, for the preservation, protection, and enhancement of the Architectural Heritage, it is essential to preserve all the values and ensure authenticity. The knowledge and preservation of the traces of time, including the current decay, patina, changes occurred over time, traditional construction techniques, evidence of natural events, and human intervention, are all parts of this history.

The well-being of the local population was considered a vital issue. The resources on which tourism is based are limited, and the demand for better environmental quality is constantly growing. These considerations have led to the idea of developing a sustainable and responsible tourism that can meet economic and environmental requirements while respecting the needs of the local population. These principles were defined during the thirty-year World Conference on Sustainable Tourism held in Lanzarote (WTO World Tourism Organization 1995).

All the general plans designed for the WHS areas were therefore developed by collecting essential studies with different objectives:

- Deep understanding of the degradation phenomena to set proper guidelines for their conservation (Figs. 3.24, 3.25 and 3.26);
- Recognizing cultural and natural heritage strengths and weaknesses in the sites' surroundings and within the core and buffer zones;

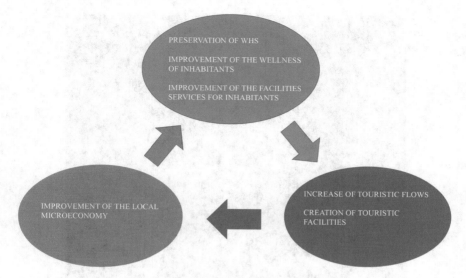

Fig. 3.22 Logical framework of the project (elaboration by authors 2022)

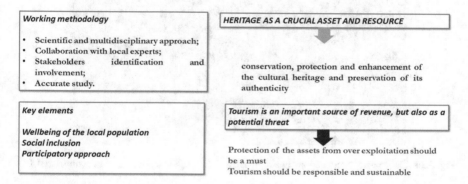

Fig. 3.23 Methodological approach (elaboration by authors 2022)

- Implementing interviews with local stakeholders to promote their involvement in tourism development with a community-driven approach;
- Identifying elements that contribute to creating the "identity of the places". These data are fundamental to planning solutions and strategies that consider the areas' specificity;
- Mapping the characters of the context to design actions able to enhance tourism development in harmony with the surrounding environment;
- Reusing and improving the existing facilities as a resource for enhancing cultural tourism.

Fig. 3.24 Diagnostic investigations in Haghpat by Nanar Kalantaryan (*Photo* by authors 2015)

This working method allowed the development of specific strategies focused on the different characteristics of the sites. The following criteria summarize the general approach detailed in the next chapter:

- Preservation of existing sites through conservation projects;
- Enhancement of the environment around the WHS;
- Protection of intangible Heritage;
- Implementation of infrastructure for the inhabitants (sewerage, water supply) with positive effects on the tourism offer;
- Recovery of traditional architecture (removal of harmful structures, preservation of historic building's facades);

Fig. 3.25 Diagnostic investigations in Haghpat by Nanar Kalantaryan (*Photo* by authors 2015)

- Creation of new facilities to diversify the tourism offer (visitor centers, laboratories, museums, picnic areas, play and fitness grounds, etc.) and increase the time visitors spend in the area;
- Involvement of the communities in the tourism market, by creating cooperati and favor widespread hospitality.

A prolonged stay in the places of interest can positively affect the local economy, triggering virtuous circles that can extend to other areas. The of innovative polarities also meets the needs of domestic tourism, which is exp develop in the coming years.

Including the WHS in a network of itineraries through the main interestin in the broader area could promote the conservation and enhancement of the ings. The study and development of different itineraries in the North wo general increase in the tourism offer. An integrated approach that consid link between Cultural Heritage and landscape could promote positive the area. The training of the local population represents another crucia process, to make them able to promote and manage new connected a pretation services, creation of Small and Medium Enterprises (SM catering, branding, marketing, art crafts, performing arts, etc.

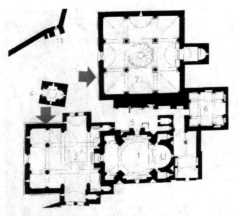

...ve

...icro
...er of
...cted to

...g points
...surround
...ld allow
...rs a be
...arketin
...aspect
...tivities
...Es), hos

...vations in Haghpat. Endoscopy tests to verify the consistency of the
...t showed the erosion of the mortar in the observed points. The test
...MArcH, Techniques for the Conservation and Management of
...o di Milano (*Photo* by authors 2015)

...i di Architettura armena. Oemme Edizioni, Venezia
...erouk. Ares, Milano
...oravank. Oemme Edizioni, Venezia

Atoyan K (2016) Tourism industry in Armenia: evaluation and perspectives. Int J Hum Manag Sci (IJHMS) 4(1):5–10. https://hal.archives-ouvertes.fr/hal-03284216/document. Accessed 9 Jul 2022

Boriani M, Premoli F (2012) Rapporto sulla conservazione, valorizzazione e gestione dei Beni Culturali della Repubblica d'Armenia, research report

Casnati G, Tonghini C (2012) Lo sviluppo costruttivo della basilica di Ererouk (Armenia), secoli VI – X: una ri-lettura archeologica. Arqueología De La Arquitectura 9:31–57

Casnati G (2014) Monuments, conservation and sustainability: the Armenian cultural, geographic and historical context. In: Casnati G (ed) (2014) The Politecnico di Milano in Armenia. An Italian Ministry of Foreign Affairs project for Restoration Training and Support to Local Institutions for the Preservation and Conservation of Armenian Heritage. Oemme Edizioni, Venezia

Council of Europe (2005) Convention on the Value of Cultural Heritage for Society, Faro, 27 Oct 2005. https://rm.coe.int/1680083746. Accessed 10 Jun 2022

Cuneo P (1988) Architettura armena: dal quarto al diciannovesimo secolo. De Luca, Roma

Harouthiounian V (1975) Monuments d'Arménie: de la préhistoire au 17e siècle A.D. Monuments of Armenia from the prehistoric era to the 17th century B.C. Société Techno-Presse Moderne, Beirut

Harutyunyan T (2017) A guide to Soviet modernisn in Yerevan. https://strelkamag.com/en/article/yerevan-modernism. Accessed 16 Jun 2022

Harutyunyan T (2018) Yerevan: architectural guide. DOM Publishers, Berlin

ICOMOS (1976) International Cultural Tourism Committee (ICTC). Icomos Cultural Tourism Charter. https://www.icomosictc.org/p/1976-icomos-cultural-tourism-charter.html. Accessed 20 Jun 2022

ICOMOS (2002) International Cultural Tourism Committee (ICTC). International Cultural Tourism Charter. https://www.icomosictc.org/p/international-cultural-tourism-charter.html. Accessed 21 Jun 2022

ICOMOS (2021) International Cultural Tourism Committee (ICTC). International Charter for Cultural Heritage Tourism. Reinforcing cultural heritage protection and community resilience through responsible and sustainable tourism management. https://www.icomosictc.org/p/charter-renewal-process.html. Accessed 21 Jun 2022

Marouti A (2018) Preservation of the architectural heritage of Armenia. A history of its evolution from the perspective of the early 19th century European travelers to the scientific preservation of the Soviet period. PhD thesis, PhD program in Preservation of Architectural Heritage, Politecnico di Milano, supervisor M. Boriani

Ministry of Economy of the Republic of Armenia (2020) Tourism development plan. https://www.mineconomy.am/en/page/89. Accessed 9 Jul 2022

Petrosyan S, Bădescu G (2016) Armenian cultural territorial system first experience. In: Rotondo F et al (eds) Cultural territorial systems. Springer International Publishing, Basel

Uppink L, Soshkin M (2022) WEF World Economic Forum, Travel & Tourism Development Index 2021. Rebuilding for a sustainable and resilient future, insight report. https://www3.wefrum.org/docs/WEF_Travel_Tourism_Development_2021.pdf. Accessed 9 Jul 2022

World Travel and Tourism Council WTTC (2018) Travel and tourism. Global economic impact & issues 2018. https://dossierturismo.files.wordpress.com/2018/03/wttc-global-economic-impact-and-issues-2018-eng.pdf. Accessed 16 Jun 2022

WTO World Tourism Organization (1995) Charter for sustainable tourism, UNWTO declarations. 5(4).UNWTO, Madrid. https://doi.org/10.18111/unwtodeclarations. Accessed 10 Jun 2022

WTO World Tourism Organization (2022) Yearbook of tourism statistics dataset [Electronic]. UNWTO, Madrid. https://www.e-unwto.org/action/doSearch?ConceptID=2202&target=topic. Accessed 15 Jun 2022

Chapter 4
Monasteries of Haghpat and Sanahin, Monastery of Geghard, and the Upper Azat. The Current Situation of the Two World Heritage Sites

Abstract This chapter illustrates the current situation of three Monasteries, with particular reference to their state of conservation. The main problems of building degradation are highlighted, starting from a short review of the principal historical events and the current consistency of the complexes. This knowledge process represents a fundamental step to addressing future interventions correctly. The relationship with the surrounding territories is no less critical for the future of the WHS, especially in terms of enhancement. In particular, the villages present many potentialities but different criticalities in all the cases analyzed. Structural measures to improve the living conditions of the resident population are particularly urgent.

Keywords Haghpat · Sanahin · Geghard monasteries · Decay survey · Surrounding villages problems and potentialities

4.1 Legal Framework and UNESCO Protection Zones

The perimeters defined by UNESCO for protecting the Monasteries of Haghpat, Sanahin, and Geghard include vast areas of the surrounding territory.

In Haghpat (Fig. 4.1), the *core zone* covers 0.7 ha, while the *buffer zone* is 8 ha. Both are enclosed in a large *building regulation zone* corresponding to the built area.

In Sanahin (Fig. 4.2), the *core zone* is 1.7 ha. In comparison, the *buffer zone* is 15.8 ha, and the *building regulation zone* includes some built rural areas around the Monastery. It widens to the river in correspondence with the bridge connecting the village to Alaverdi. A protective area of 0.2 ha surrounds the bridge.

Lastly, in Geghard (Fig. 4.3), the protected areas include the *core zone* around the monument and a *buffer zone* extending to the Upper Azat Valley (UNESCO).

There is a substantial coincidence between the zoning set by UNESCO and the areas of protection (*bahbanagan kodi*) defined in 1989 by the Ministry of Culture of the Republic of Armenia. For Haghpat (Fig. 4.4) and Sanahin (Fig. 4.5), farming activities are permitted in the protected areas, provided they do not affect the landscape features. The buildings should not exceed 7 m in height (Haghpat) or 7.6 m (Sanahin).

M. Giambruno and S. Pistidda, *Heritage for a Sustainable Development: The World Heritage Sites and Their Impacts on Cultural Territories*, PoliMI SpringerBriefs, https://doi.org/10.1007/978-3-031-20157-8_4

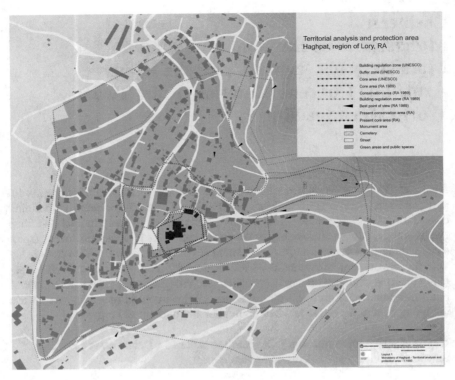

Fig. 4.1 Monastery of Haghpat, territorial analysis and protection area (elaboration by authors)

The studies, the archaeological excavations and all the interventions on the monasteries must be performed by expert scholars. The works should be authorized by the Ministry of Culture, as well as any intervention on buildings located within the perimeter. The protection regulation also defines some perspectives on the sites to protect.

In Geghard (Fig. 4.6), the protection zone includes 27 hectares of land and safeguards the site and the natural landscape in its surroundings.

For all three sites, some variations have been recently introduced to the perimeters of the protection areas defined by the Ministry of Culture (core zones), for them to follow the boundaries of the private properties.

As for city planning, the situation is very different in each of the three sites.

Only Sanahin, as a district of Alaverdi, is included in a development plan that defines the zoning of the territory. Apparently, no detailed planning tool for the historical centers and/or buildings of cultural interest has been prepared.

The situation appears quite restrictive and controlled from the point of view of national protection and the definition of the WHS perimeters. However, the total absence of constraint policies on the historical urban landscape surrounding the three monasteries could lead to profound changes that would impact not only the sacred

Fig. 4.2 Monastery of Sanahin, territorial analysis and protection area (elaboration by authors)

complexes but, more generally, the preservation of the architectural and landscape heritage of the three places. The only indications of protection concern perspective views and the height of the buildings. There is no indication of the need to preserve the existing historic buildings, the full-empty relationship that characterizes the villages, or the link between the built environment and the three Monasteries.

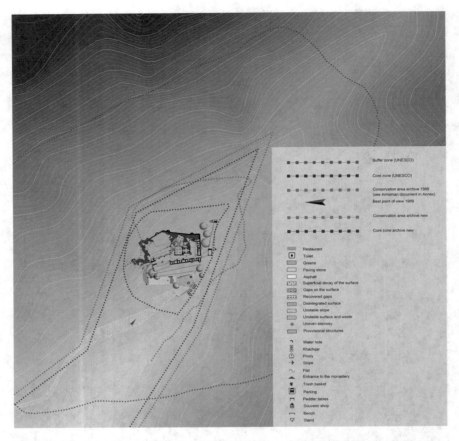

Fig. 4.3 Monastery of Geghard, territorial analysis and protection area (elaboration by authors)

Until now, the non-thriving economic conditions of the inhabitants have preserved the area due to the absence of interventions. However, some signs of possible replacement of historical buildings can already be present.

The design and implementation of a detailed urban plan for the WHS areas are necessary to favor their sustainable development as well as the definition of criteria and guidelines for guaranteeing the safeguard of the architectural heritage and historic landscape of the WHS.

The maintenance of the current ratio of construction and the territorial building index is another critical element. Guidelines for preserving the historic buildings in the villages would also be essential to avoid uncontrolled transformations, even in self-construction.

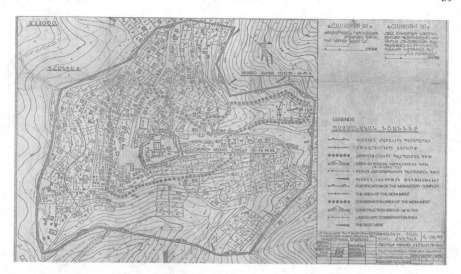

Fig. 4.4 Monastery of Haghpat, areas of protection (*bahbanagan kodì*) defined in 1989 by the Ministry of Culture (Archive of Ministry of Culture of the Republic of Armenia)

Fig. 4.5 Monastery of Sanahin, areas of protection (*bahbanagan kodì*) defined in 1989 by the Ministry of Culture (Archive of Ministry of Culture of the Republic of Armenia)

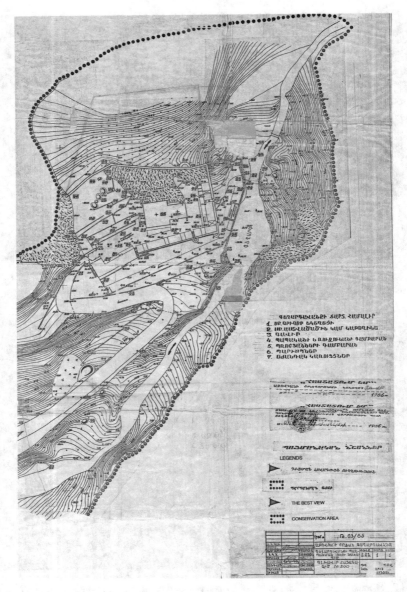

Fig. 4.6 Monastery of Geghard, areas of protection (*bahbanagan kodì*) defined in 1989 by the Ministry of Culture (Archive of Ministry of Culture of the Republic of Armenia)

4.2 Haghpat

4.2.1 Historical Framework and Current Degradation Problems of the Monastic Complex

Haghpat monastic complex was built in the tenth century on a *plateau* at an altitude of about 1000 m, 10 km from the town of Alaverdi.

It consists of a big church devoted to St. Nshan (Holy Cross, 966/991), with a vast adjacent *gavit* built during the second half of the 13th c., two small churches (St. Gregory—1005/1025 and St. Mary—13th c.), the "building of Hamazasp" (built during 13th c.) linked to the library building (built at the beginning of 12th c. and rehabilitated several times) by a large hall (Fig. 4.7).

Fig. 4.7 Monastery of Haghpat, general plan (elaboration by authors)

Fig. 4.8 Monastery of Haghpat, top view (*Photo* Giambruno 2015)

The famous *khatchkar* of St. Saviour (1279), one of the very few *khatchkars* in Armenia representing the crucifixion, is conserved in the hall. On the North, the impressive three-storied bell tower (1245) was built in the upper part of the complex.

Relatively isolated, in the North-East is the refectory (13th c.) and in the South East are the tombs of Oukanants (first half of the 13th c.) supporting three big *khatchkars*.

The whole complex is surrounded by walls and watch towers that varied during the centuries as the defense concepts changed (Figs. 4.8 and 4.9).

Climbing towards East, about 100 m away from the enceinte, there is a roofed fountain built in 1258 (Fig. 4.10), and another 200 m up is a chapel devoted to the Virgin Mary (12–13th c.), flanked by wonderful *khatchkars*.

The Monastery was a significant cultural center whose life was strictly connected with the adjacent village life (Fig. 4.11) (Armenian monks used the what is known as the idiorhythmic system, in which the monks used to live in the village but had communal life was only confined to the religious functions).

Its architecture was continuously adapted to contingent needs, thus resulting in several restoration works undergone since the beginning.

During the 11th and 12th c., the Church of St. Nshan was restored (probably the row of decorated ashlar stones near the top of the dome was added during the 12th c.). In the 13th c., the domed Church of St. Gregory was transformed into a barrel-vaulted building.

Fig. 4.9 Monastery of Haghpat, general view (*Photo* Giambruno 2015)

In the 18th c., several restoration works were implemented on various buildings, and presumably in the early 1800s some rehabilitation activities occurred to adapt the Monastery to serve as Holy See under Eprem Catolicos (AA VV 1970, Ghirimezi n.d., Jean de Crimè 1863).

The complex now presents various deterioration and instability phenomena that should be carefully considered.

The roofing is overgrown with weeds, and sometimes even by small trees, that trigger micro fracturing processes extending and expanding gradually due to the repetition of freezing and thawing cycles, very frequent in the Armenian climate (Fig. 4.12).

Some roofing tiles slide and protrude by about 10/15 cm (for example, just above the *gavit* entrance).

Local witnesses report that this phenomenon increases at every rainfall, which demonstrates the lack of connection between the tiles and their support, the roofing instability and the consequent need for urgent interventions.

During some on-site works, an inspection performed by climbing on the roof has proved that the tiles are nearly entirely detached from their support, which allows the penetration of soil and the utilization of the voids by different animals. Furthermore, a survey realized after a rainfall permitted us to verify the conspicuous water penetration from the roofs, mainly in the *gavit*. The care keeper of the complex reports

Fig. 4.10 Haghpat, the fountain around the Monastery (*Photo* Giambruno 2015)

that, during an operation of tile restoration, he realized the absence of mortar in many parts of the coverings.

This situation is common to all the sites object of the study and other Armenian monuments investigated in the past by the Politecnico di Milano.

This wake-up call demonstrates the need to improve the technique adopted to restore the roofs, in order to guarantee a better and long-lasting connection between the tiles and the under layer.

Regarding the masonry, some of the ashlars are rotated and leading out (Fig. 4.13). In many areas, the internal mortars have probably lost their binder material, as highlighted by the presence of wall surfaces affected by the stagnation of humidity, concretions and salts. An endoscopic survey has shown that in the interior of the walls, the mortar is disintegrated in many points.

Fig. 4.11 Village of Haghpat (*Photo* Giambruno 2015)

Major cracks are found on the outside wall of the bell tower and in the interior of the Church of St. Nshan. On both the interior and the external surfaces of the buildings, we can find thick whitish deposits (Fig. 4.14). The nature of these deposits has been determined through laboratory analysis of samples especially taken and studied at the Department of Civil and Environmental Engineering of the Politecnico di Milano. The investigations showed that the deposit mainly consists of calcium carbonate; this demonstrates the presence of soluble salts due to the surface percolation of rainwater.

4.2.2 Problems and Potentialities of Haghpat Village

Haghpat is one of the 27 municipalities of the Tumanyan province in the Lori Region, in the North of Armenia, whose origin dates back to the eighth—seventh centuries BC. It is 178 km far from Yerevan (about three and a half hours' drive). The connection takes place through two different routes, one in the West crossing Aparan and the other, only 10 km longer, through the city of Sevan, a famous and well-developed tourism destination that mainly attracts domestic tourists and tourists from the CIS countries. The municipality's surface area is 15.28 km^2, and the village is made up of 270 houses that host 767 inhabitants.

Fig. 4.12 Monastery of Haghpat, decay phenomena on the roof (*Photo* Giambruno 2015)

The name of the village derives from the name of the Monastery. The original was Haghbat (*haghb = at trap and at = hate*) which later became Haghpat.

The village is built on a pleasant highland with deep valleys and steep slopes, 10 km far from Alaverdi, an important industrial center linked to the mines of molybdenum, dramatically polluted.

The village surrounds the Monastery and has some exciting buildings. They are all pretty recent (only a few houses date back to the last century), and were realized on the ruins of previous ones. According to the literary sources, the previous village had to be quite large, and hundreds of monks lived there during the years of the Monastery's splendor. In general, the buildings are well inserted into the landscape and feature characteristic elements of the traditional architecture suitable to form the backdrop of the Monastery (Brady Kiesling 2001).

On the West, the village is surrounded by deep canyons, while on the East side, the Surblis and Terunakan mountains protected Haghpat from invaders and now offer rich pastures (the villagers live mainly from agriculture and livestock breeding).

Part of the products from these activities (butter, milk and, cheese) serve the tourism industry as they are sold to hotels and restaurants. About 40 people from Haghpat work in the Teghut mine for ACP (Armenian Copper Program).

Haghpat was named "the red" for the color of its brick tiles roofs and for the presence here of the first Armenian Bolsheviks.

Fig. 4.13 Monastery of Haghpat, decay phenomena on the masonry blocks with rotation of the ashlars (*Photo* Giambruno 2015)

A lovely 19th c. building with a wooden balcony is conserved next to the fountain, 100 m out of the monastery walls (Fig. 4.15). In the past, the building was the Stepan Shahumyan Bolshevik social club and is now the property of the Communist Party. Today it is abandoned and the Mayor has recently expressed the will to regain the house museum to the community, to allow its opening to visitors.

The village has a kindergarten and a school with computers and internet access. Haghpat hosts a musical school and a library, both affected by the lack of heating.

A new chess center/school is opening in a few weeks, and an internet point with five computer stations will be realized to give internet access to the villagers who cannot have it at home. The Municipality's Cultural center is closed, but a concert hall is available near the City Hall and hosts every kind of spectacle and meeting, offering about 200 seats.

From the infrastructure point of view (source: the Mayor), the village is not served by running water which, therefore, is not available in all the buildings. Similarly, there is no sewage system; the buildings are equipped with self-discharge systems and consequently, many houses have external toilets in the green areas.

The main roads are paved, though they would require maintenance or the realization of new paving; the secondary routes serving many homes are made of clay.

Fig. 4.14 Monastery of Haghpat, decay phenomena on the exterior façade, with black crusts and efflorescence (*Photo* Giambruno 2015)

The public space in front of the accesses of the monastic complex (sidewalks, fences) need general redesigning (Fig. 4.16).

A quick factsheet has been drafted for the village's buildings, including the WHS buffer zone, precisely aimed to define the general state of conservation, the materials and decay of the roofs, the ownership, the current use and the presence of significant elements of traditional architecture (Fig. 4.17).

The collection of this information was of great use in identifying critical housing conditions and directing priorities for action to improve the inhabitants' quality of life, a factor of great importance in itself and a fundamental prerequisite for the development of tourism in the area.

Among the 58 buildings studied, 48 have roofs in asbestos, four are partially used and 13 are unused. Regarding the conservation state, two are in ruins and 8 have structural problems. While only a small number of buildings are in severe dilapidated conditions, the presence of asbestos in the dwellings is a very serious danger (Fig. 4.18).

Tourist accommodation facilities are limited to some rooms in the building facing the access square; a guest house is under construction a few meters from the Monastery on the way to the City Hall.

Some interviews with the local people and visits in situ have allowed a first identification of the families willing to transform their houses into guesthouses or

Fig. 4.15 Village of Haghpat, Stepan Shahumyan house (*Photo* Giambruno 2015)

B&Bs. Some rooms dedicated to this use are already available in a few residential buildings. On the other hand, there are no reception points or tourist information centers, and there is no explicative or directional signage for the village and the WHS.

The open spaces near the Monastery do not offer facilities to stop and rest. Souvenirs are sold in a few kiosks in precarious conditions, facing the main square and hiding a little Medieval chapel; car parking is allowed with no regulation on the square in front of the religious site.

4.3 Sanahin

4.3.1 The Monastic Complex. Historical Overview and Today's Problems of Decay

The Sanahin monastic complex, built on an earlier church of unknown dedication, has been mentioned in medieval literature since the beginning of the tenth century. In contrast, in 979 it was selected as the See for the bishop of the reign of Tashir-Dzoragued, thus becoming a great cultural center. The historian Stephen of Taron

Fig. 4.16 Haghpat, the entrance square to the Monastery (*Photo* Giambruno 2015)

reports that around 500 monks lived and worked in Haghpat and Sanahin from the beginning of the tenth century; likewise, painting, medicine and miniature were taught in Sanahin at the end of the tenth century.

In 1235, the Mongol invasion marked the start of decadence. Many monuments were lost in that period: St. James church, the 10th c. *gavit*, all the monks' dwellings, the Kiurikian tomb, the 18th c. caravanserai. At the same time, the frescoes and the decorations in the interior of remained buildings were strongly damaged.

The complex currently consists of the following monuments: the St. Asdvatzadzin church (St. Mother of God, built-in 934) with an adjunct *gavit* (1211); the bigger Church of St. Amenaprkitch (St. Saviour, built-in 966), also with an adjunct *gavit* (1181); the academy (built in two steps from the beginning and the end of the 11th c.); a round chapel devoted to St. Gregory (1061); a Bibliotheque (1063) fronted by a porch (18th c.) and a belfry (12th-18th c.); the remains of the Kiurikian tomb; and, a little distant, the tomb of the Zakaryan family. The complex is included in an enceinte made of stone walls (Fig. 4.19). In front of the 1212 *gavit* is a famous cross-stone (*khatchkar*) engraved by Mkhitar in 1184 (Ghirimezi n.d., Jean de Crimè 1863).

Due to its troubled history and the occurrence of earthquakes, the building we can study today is the result of several interventions. In particular, we know from the literature that in 1181 the abbot Hovhannes promoted rebuilding the *gavit* in front of St. Amenaprkitch church and the restoration of the Church with the reconstruction of the cupola, which had collapsed due to an earthquake.

HAGHPAT	Inventory number: *H-20*

PHOTO

1.GENERAL DATA

1.1.Period of construction: Unknown	1.2.Typology: Residential	1.3. Ownership (Public /Private) Private

1.4. Stories of the building: One story

2. GEOGRAPHIC LOCATION

2.1. Urban context	2.2.Address:
	2.3. Road accessibility: Accessible by Car way and by pedestrian way
	2.4. Demographic situation: **Used** ☒ **Partly used** (dwellers mainly live in the city and only in the summer move to the village) ☐ **Abandoned** ☐

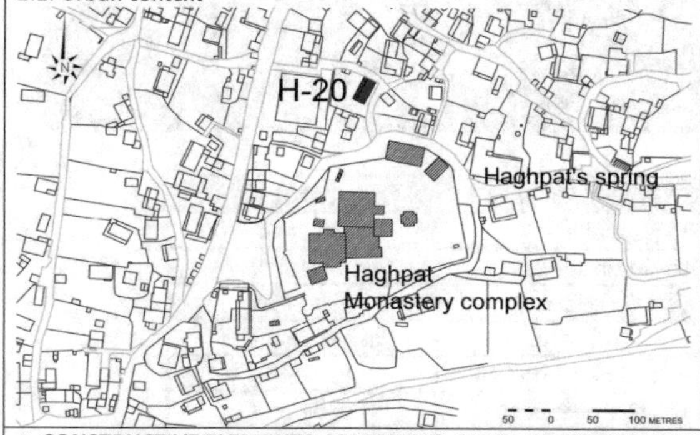

3. CONSTRUCTIVE ELEMENTS, MATERIALS, STATE OF CONSERVATION AND INTERVENTIONS

a. Materials and characters of facades Main material of the building is tuff **State of conservation:** Satisfied, with superficial decays **Interventions:**	b. Roofs: Gable roof Half asbestos and half ceramic tiles **State of conservation:** Satisfied, with superficial decays **Intervention:**

PROBLEMS

NOTES
The building has wooden balcony

Fig. 4.17 Haghpat, an example of factsheet for collect information on the village's buildings (elaboration by authors)

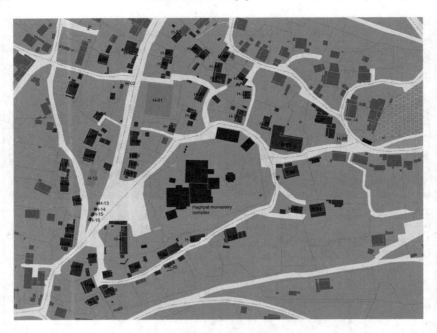

N	USE	GENERAL DECAY STATE	ROOF
H-01	used	no degradation	metalic
H-02	used	no degradation	metalic
H-03	used	no degradation	asbestos
H-04	unused	ruin	asbestos/tiles
H-05	partly used	superficial degradation	asbestos
H-06	used	Localized collapses	asbestos
H-07	used	no degradation	asbestos
H-08	used	superficial degradation	asbestos
H-09	used	no degradation	metalic
H-10	used	superficial degradation	asbestos
H-11	unused	structural instability	asbestos
H-12	unused	superficial degradation	asbestos
H-13	unused	structural instability	
H-14	unused	superficial degradation	
H-15	used	no degradation	metalic
H-16	used	superficial degradation	asbestos
H-17	used	no degradation	metalic
H-18	partly used	Localized collapses	asbestos/tiles
H-19	partly used	superficial degradation	asbestos
H-20	used	superficial degradation	asbestos/tiles
H-21	used	superficial degradation	asbestos

Figs. 4.18 Haghpat, an example of thematic map about the use of the buildings, the state of conservation and the roof condition (elaboration by authors)

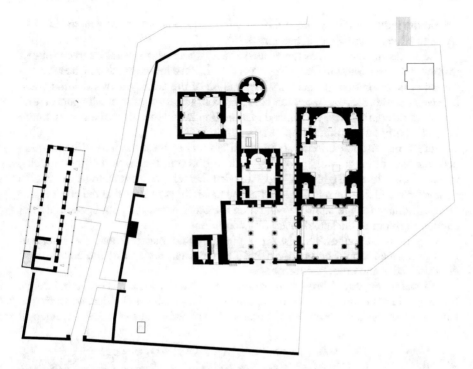

Fig. 4.19 Monastery of Sanahin, general plan (elaboration by authors)

In 1652 major restoration works were undertaken to rebuild what was destroyed by a strong earthquake. The drum and the cupola of the Church St. Asdvatzadzin were rebuilt, together with all the roofing which was built higher. Also the façade of St. Amenaprkitch was repaired at that time, together with the St. Gregory church whose cupola was restored several times.

Today's aspect of the St. Gregory church is the result of several restoration interventions that have reduced the external volume compared to the original one.

We also have information about the fact that in 1815 and 1881, some other restoration works were undertaken, but more detailed descriptions of their consistency has not been found so far, apart from a picture dating back to the beginning of the same century where the drums and dome of the belfry, the churches, and the St. Amenaprkitch *gavit* appear as they were brand new.

From 1969 to 1990, the archive registers contain 43 restoration projects for the preservation of the *gavit*, the Church of St. Grigor, and the preservation of St. Astvatsatsin and St. Amenaprkitch churches.

Just in front of the 1211 *gavit*'s facade lies a big and rusty metallic structure for uplifting stones (presumably for restoration works). As we noticed from previous visits to the site, it was installed (but not working) in 1987, while it was not visible in the picture published in *Documenti di Architettura Armena n°3 Sanahin* dated 1980 (Alpago Novello and Ghalpakhtchian 1970).

The monastic complex, as mentioned above, underwent several interventions of restoration/reconstruction that make it very difficult to interpret the consistency and causes of deterioration, particularly the reading of the crack pattern. A wide crack is present in the St. Amenaprkitch church. From observing the cracks pattern and structural discontinuities, we noticed that the Southeastern wall of the main Church is significantly out of plumb (Fig. 4.20).

Here some disconnections can be identified that suggest the lack of connection between the different walls, probably built at different times. Some photographs, mainly dating to the Eighties of the Twentieth century, show a state of the cracks comparable with the current one. On the one hand, this suggests that the cracks have been stationary for the last forty years; on the other hand, the data is certainly not enough to interpret the actual instability conditions.

A systematization of all historical photographic material and the interventions performed would be greatly helpful to the interpretation of the cracks, wall discontinuities and other phenomena.

The water penetrates through the roofing in the vaults and walls of the buildings. It leaves whitish deposits (calcium carbonate, soluble salts, and silica, according to laboratory analysis done by the Laboratorio Prove Materiali of the Politecnico di

Fig. 4.20 Monastery of Sanahin, decay phenomena on the façade: walls discontinuity, staining and vegetation on the roof (*Photo* Giambruno 2015)

Fig. 4.21 Monastery of Sanahin, decay phenomena due to water infiltration (*Photo* Giambruno)

Milano) that could also represent the soiling of cement used in past restoration work (Fig. 4.21).

Some roofs have been recently restored, but we could not collect any information about the techniques and materials adopted. It should also be helpful for future reference to monitor the efficacy of these interventions verifying if the underlying walls are dry.

The weatherboards, maybe due to their incorrect shape (too short), convey water directly on the walls. The roofing is overgrown by weeds and, where not restored, also by bushes and little trees; in some parts, the tiles and a few stone ashlars are missing or moved from their original setting (the roof of St. Amenaprkitch church presents some tiles sliding towards North).

The presence of weeds is testified, differently from Haghpat, also on some walls, particularly on the Northern facade of St. Amenaprkitch. The roots, in some cases quite big, can cause severe fractures in the masonry.

A comment is needed concerning the works of replacement/installation of new stone slabs between the tombstones in the interior of the *gavit*. This intervention should be better designed to be more compatible with the historical materiality of the complex. A picture of the Seventies, shot by Alpago Novello, shows the interior of the Church St. Asdvatzadzin with the remaining wall paintings, no longer existing, and a wooden iconostasis in front of the altar of which no traces remain.

4.3.2 The Problematic Relationship Between the Sanahin Settlement and the Monastery

Sanahin is not considered a village but a neighborhood, which implies significant differences in its inhabitants' way of life and resources and development possibilities. Like Haghpat and Geghard, the Sanahin monastery is owned by the Holy See of Etchmiadzin (since April 13th, 2011) but, strange enough, nobody in the surroundings appears to know that the local population is complaining about the lack of a priest to serve the Church.

The Sanahin neighborhood has been recently annexed to the nearby Alaverdi municipality and two areas can be distinguished in it based on architecture and construction features.

The first zone, distant from the Monastery, has urban features and is characterized by buildings realized during the Soviet era. Its central square, with "monumental" soviet buildings, displays a quite interesting urban design, although deteriorated. Nearby is the arrival station of the ropeway connecting Alaverdi to the Sanahin highland (Fig. 4.22). The arrival station construction is exciting; it is situated on a little plateau that dominates the underlying town.

The part of the village near the Monastery has a rural character with low buildings. Some are a compelling testimony of traditional rural architecture, and are surrounded by green spaces devoted to harvesting products for own consumption or the local market (butter, yogurt, cheese, etc.) (Fig. 4.23).

The study concerning the portion of the village included in the buffer zone was realized by compiling factsheets which outcomes were applied on a map to obtain a straightforward interpretation of the results. The survey included 68 buildings, of which 42 with asbestos roofing.

Two buildings are in ruins, while three show severe structural problems. As in Haghpat, also in Sanahin a certain number of buildings are not used (10) or partially in use (7).

The water mains reach the neighborhood, but the sewage system has not yet been realized. Public lighting and roads need improvement.

The streets, often uneven, do not have sidewalks.

The small square facing the Monastery includes all the services designed for tourism (Fig. 4.24). There are souvenir shops (provisional kiosks and canopies provide repair for the peddlers placed along the stairway that leads to the Monastery), parking lots, and toilets, now arranged in a new building set in an improper area.

The imposing former Sanahin Café is placed between the Monastery and the road from which one arrives (Fig. 4.25). The building presents structural and material decay together with a partial collapse of the roofing (realized in asbestos). The structure is an integral part of the complex and a sign of its recent history. A conservation intervention would allow the reuse of both the two floors of the building. Currently, only the upper floor is used, as a memorial hall for funeral ceremonies.

Fig. 4.22 Arrival station of the ropeway connecting Sanahin to Alaverdi (*Photo* Giambruno 2015)

In the nearby Monastery, we noticed the lack of structures intended to host tourists. Some interviews to the inhabitants have shown that seven families are willing to establish a B&B, but due to financial constraints, they need adequate supported.

At a short distance from the Monastery, the house museum of the Mikoiyan brothers is found (Fig. 4.26).

They were born in Sanahin; Anastas Mikoyan was a member of the Soviet Politburo, while his brother Artem was an aeronautical engineer. Together with Gurevich, they designed the famous MIG. The two figures of the Mikoiyan brothers (the politician and the aeronautical engineer) lend themselves to urge visitors' curiosity, recreating the atmosphere of the communist era, the political infighting, and the Soviet military technology.

Fig. 4.23   A building in the village near the Monastery (*Photo* Giambruno 2015)

This little house museum can offer an alternative itinerary for the visitors, increasing their time spent in the place and widening their experience. The Mikoiyan house is preceded by a beautiful open space decorated with a fascinating monument of the soviet times, including a MIG 21, one of the military planes designed by Artem Mikoiyan. The area is pleasant and solemn. A walking path flanked by high boxwood bushes leads to the house museum.

4.4   Geghard

4.4.1   Degradation and Disruption of the Closest Monastery to Yerevan

Approximately 40 km from Yerevan, at the end of a narrow valley and surrounded by mountains, the Geghard Monastery owes its uniqueness to the coexistence of rooms built and rooms carved into the rock.

The origins of the complex are uncertain, although some sources date it to the roots of Christianity, according to some 7th c. inscriptions found on its territory. The first historical reference dates to the tenth century and tells about the destruction of

Fig. 4.24 The square at the entrance of the Monastery (*Photo* Giambruno 2015)

Fig. 4.25 The Sanahin Café near the Monastery (*Photo* Giambruno 2015)

Fig. 4.26 The entrance to the Mikoiyan museum (*Photo* Giambruno 2015)

the Monastery (named Ayrivank—Monastery of the cave) by the Arabs. The oldest part of the actual Monastery is the little chapel half-built and half-carved in the rocks, dedicated to St. Gregory (previously St. Asdvatzadzin), built in the 12th c.; the current layout of the site is also dated to the twelfth century.

The existing monastery complex is said to derive its name from the lance which pierced the side of Jesus and whose relics were conserved therein (they are now in the Museum of the Holy See in Etchmiadzin).

The site has been repeatedly remodeled and reconstructed in some of its parts, including recently. It consists of the main Church (*katoghiké*—1215), a *gavit* (entrance hall—1225), two churches carved in the rock (18th c.), and a *jamadun* (*gavit* cemetery—1283) (Fig. 4.27).

The Monastery is surrounded by a system of walls with semicircular towers that were restored and modified during the centuries. Lining these walls on the East and South are the quarters intended for civilian use (17th c.). There are two entrances, one on the West side and the other on the East side. In the enclosure and on the mountain's rocky slopes are numerous cells and chapels carved in the rock, even on several levels (Fig. 4.28).

At the North East of the complex, about a hundred meters above the Monastery, the remains are found of a group of cells dug into the rock overlooking the valley (Fig. 4.29).

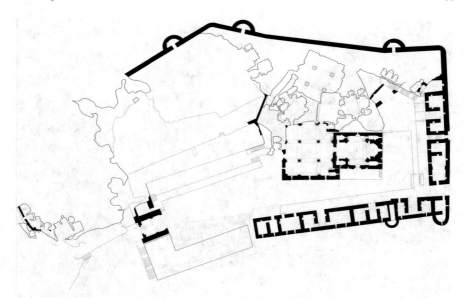

Fig. 4.27 Monastery of Geghard, general plan (elaboration by authors)

Fig. 4.28 Monastery of Geghard, general view from the access road (*Photo* Giambruno 2015)

Fig. 4.29 Monastery of Geghard, the churches dug in the rock (*Photo* Pistidda 2022)

Interestingly enough, North West, in a predominant position on the mountain, outside the monastery walls, there was a huge chamber entirely dug out of the rock, with a rectangular plan of 9 × 15 m and 5 m high. The chamber was excavated in 1932 by architect Toramanyan and, left open, collapsed in 1967, leaving a large hole in the mountain that is now dangerous and should be somehow protected and consolidated to avoid risks for the visitors. All around the site are many *khatchkars* carved in the rocks (Fig. 4.30) (Sahinian 1958; Sahinian et al. 1973).

The first rock-carved church hosts a natural spring, worshipped in ancient times and today conserved and respected. The presence of natural water in the rocks may also has a downside, if it is not well canalized, as it can increase the natural processes of rock deterioration.

Over the centuries, this phenomenon was kept controlled and the Monastery underwent several restoration works that should be studied with due attention.

The Geghard's monastic complex, compared with Haghpat and Sanahin, is the one that is less suffering from deterioration. However, some heavy reconstructions, particularly the surrounding walls and civil buildings, have partially falsified their authenticity.

The recent interventions, particularly for the rainwater drainage from the mountain above it, have probably solved or hidden some of the more evident problems.

Furthermore, the daily use of the buildings allows continuous maintenance work that makes a big difference.

The roofing of the Church and *gavit* is not older than 50 years (in the Seventies, they were covered by a metallic shed, as testified by the pictures shot by Alpago

Fig. 4.30 Monastery of Geghard, the churches dug in the rock (*Photo* Giambruno 2015)

Novello). Nevertheless, it already faced severe water penetration problems, only partially solved last year thanks to the intervention financed by the Ministry of Culture of the Republic of Armenia.

As far as the enclosure and lining buildings are concerned, we recommend paying attention to avoid an excess of reconstruction work that may falsify the complex's historical consistency, leading to the loss of its value as a historical testimony and of its "charm" (Fig. 4.31).

The presence of whitish deposits, similar to the ones found at Haghpat and Sanahin, suggests a problem of water penetration from the roofing (past or still ongoing), and a lack of binding properties of the mortars in the internal part of the three-layered walls of midis type masonry. This is particularly true for the drum.

A survey was undertaken just after a snowfall and another during a rainy day which allowed to notice the presence of water on the ground. In this case, the situation is much more complex as many parts of the monastery are hypogea, directly carved in the bedrock. For the same reason, in these parts, the crack pattern is difficult to interpret as it involves the bedrock in which many components of the Monastery are realized (Fig. 4.32).

The cracks affect vaults, walls, and elements of the carved parts of the complex. A comparison with some pictures dating back to the Seventies and Eighties of the twentieth century shows a crack pattern comparable with the current one, thus demonstrating that, when active, it is a slow process.

Fig. 4.31 Monastery of Geghard, ongoing works on the main building (*Photo* Pistidda 2022)

There is a biological patina on the areas affected by more extensive water infiltration, while on the surfaces of the interior and exterior spaces deposits of soluble salts are found, always due to storm water infiltration. A recent restoration project, of which the MoC (Ministry of Culture) has provided a copy, includes the construction of a drainage and flood control system running from the mountain slope to the Monastery (already realized) and the sealing of the cracks with injections (not yet implemented).

Greater attention should be devoted to the monuments located outside the enclosure: in spite of their being of great historical interest, they are abandoned and seriously damaged. For this part, as for the whole site, it is of utmost importance to acquire all the archaeological documentation and any pre-existing geological studies, or commission new studies.

4.4.2   The Arrival Places and the West and East Areas Around the Monastery

The large square giving access to the Monastery is a large asphalt ribbon, just as the broad road leading to the complex (Fig. 4.33).

Vehicle parking is not regulated, and in the vicinity, the Azat riverbank presents small landslides in several points and displays a lack of land coherence due to erosion during floods (the geological map shows that this is a pretty widespread phenomenon

Fig. 4.32 Monastery of Geghard, a big crack on the entranche of the Monastery (*Photo* Pistidda 2022)

in the area) (Fig. 4.34). It should also be noticed that the area is polluted by the rubbish (bottles, plastic bags, etc.) carelessly left by visitors.

On the land slope just under the road, illegally installed metallic structures with provisional sheds are found that intended for picnics but incompatible with UNESCO's requirements for the core zone.

The only positive observation regards the little spontaneous market which has a long history. It consists of a row of tables decorated with taste which offer mainly food, especially dried fruits but primarily the famous cake *gatà,* the tastiest one in

Fig. 4.33 Monastery of Geghard, the parking area at the entrance (*Photo* Giambruno 2015)

Armenia, prepared every morning by the same women who sell it. It is decorated with the wording *GEGHARD* in Armenian letters, so it also has significance as a souvenir (Fig. 4.35). We cannot say one goes to Geghard specifically for having *gatà;* however, buying this cake is a must also for local tourists. This market typology does not provide any comfort to the sellers who must bring their metallic tables from home every day, but adds color and increases the interest in the place, as shown by the fact that most travel guides recommend visiting it.

Very often there is a group of musicians wearing traditional costumes and playing Armenian folk music, and tourists generally enjoy their presence.

The pedestrian path to the complex is paved with little basalt stones somewhere disconnected. On the turnabout, there is a widening in beaten earth where AMAP (The Armenian Monuments Awareness Project) has installed a large poster that presents the natural richness of the area in several languages, and a little bit uphill is the "path" climbing to the rock chapels (Fig. 4.36). A cement disconnected stair leads to a steep slope scattered with big rocks and flanked by a dirt trail that adverse weather conditions can make very slippery and dangerous both for those walking there and those passing down. The vast cave corresponding to the chamber found by Toramanyan and collapsed in the Sixties has been abandoned as it is; now it is difficult to understand what it is and visiting it could be very dangerous, as the risk of further collapses persists.

Fig. 4.34 Monastery of Geghard, the Azat riverbank with many landslides (*Photo* Giambruno 2015)

The visit to the St. Gregory chapel and other rock constructions is very charming but also dangerous, as there is no protection and no interpretation. The chapel may also be reached via a path that departs from the right side of the asphalt road, before the parking; it is initially paved with basalt stone and flanks many little caves or rooms carved in the rocks. The West entrance to the Monastery is quite imposing and adequate to the importance of the site. Once you enter, the Monastery is on the left, and on the right is a souvenir/candle shop, together with the buildings used by the monks.

On the East, one can exit the monastery enceinte through a narrow portal, after which a paved path leads to the picnic area (*khorovatz)* and the toilets for visitors, indicated by very simple and not well-designed signs. Uphill of the Monastery, there

Fig. 4.35 Monastery of Geghard, the famous cake *gatà* (*Photo* Giambruno 2015)

Fig. 4.36 Monastery of Geghard, the path to the rock chapel (*Photo* Pistidda 2022)

Fig. 4.37 Monastery of Geghard, the orchard and the honey production (*Photo* Pistidda 2022)

is a vast orchard protected by a metallic fence and intended for the use of the monks; here they also produce honey (Fig. 4.37).

Armenian visitors traditionally use the wide area on the East of the monastery wall for having a picnic after the *madagh* (a sort of sacrifice, a tradition whose origins date back to the ancient times of paganism). An ugly metallic structure just out of the exit from the Monastery is intended to be the place where animals (generally roosters or lambs) are "transformed" into skewers. A minor pedestrian bridge makes it possible to cross the river and reach the other slope where one can find a famous—although not attractive—colossal cave. There is a traditional wish tree were one can tie little pieces of fabric representing their wishes, and several pleasant spots to enjoy the shadow of the trees.

Behind the river, little trails lead to non-equipped picnic areas and, about 9 km far, to interesting and very ancient archaeological sites.

To go back to the parking lot flanking the Southern part of the enclosure one can walk on a terrace embellished by some fruit trees. The external prospect of the monastic complex appears in this portion as the result of a series of massive restoration and/or reconstruction interventions not coordinated in a comprehensive project.

During an informal interview with one of the monks living there, we learned that there is a project for enhancing the site that provides for various interventions, including the realization of some rooms for hosting pilgrims, the completion of new toilets just lining the walls enclosure, the displacement of the "historical" market.

Along the road leading to the site, a restaurant faces the steep river valley, and in front of it is a pleasant grassy slope. This area of the site, up to the St. Gregory chapel, features fractured stone blocks and landslides that make the access road risky for visitors. Some improvised picnic areas, set in charming spots, flank the river bed.

4.4.3 The Goght and Geghard Villages. When Tourism Does not Benefit the Local Population

About five kilometers away from the monastic complex, slightly downhill of the road leading to the Monastery, is Goght village, known since the thirteenth century with the name of Goghot. It is made up of 527 houses that host 2030 people, out of whom only about two dozen are currently involved in tourism activities linked to the presence of the famous Monastery. The others are working as farmers: livestock breeding is organized in medium-sized farms, but fruit and flowers cultivation is the main asset (an increase from 32 to 80 ha has been planned of tree cultivation area). In 2010 a vast greenhouse for the cultivation of roses was built that currently gives work to 150 people. Another greenhouse for cucumber cultivation employs 250 people, a number which is expected to doubled next year. Its impact on the landscape should be verified.

Many of the abovementioned assets are the result of a program financed by the Green Land Investment Company started in 2008. Although the municipality is quite wealthy, it still has no sewage system—which would require a 150 mln drams investment—and no medical assistance facilities nor a pharmacy.

The territory of the municipality (12.30 km^2) included the entire Monastery area until this was transferred to the Holy See of Etchmiadzin by a Governmental decision in 2004.

In our opinion, the inhabitants of Goght should and could be more involved in the tourism activities by creating synergies with the monastic complex and, possibly, with the Garni village, with which some positive exchanges already exist.

The village has a simple and pleasant architecture, composed of houses scattered and surrounded by fenced green areas like a typical Armenian rural village. The main square, situated in the lower part, would need some intervention to be requalified and redesigned, as the surrounding buildings also do. We learned from an interview with the village mayor that they want to transfer the town hall to another facility and allocate the civic center here. This would undoubtedly good, as the village lacks a polarity and a social and architectural reference point.

In the nearby is the Azat river, surrounded by pleasant vegetation and easily reachable through a dirt road.

A change in direction from the main road leads, in three kilometers, to the tiny hamlet of Geghard. The road climbs up to 2000 m above sea level, passing through cultivated fields and orchards.

The Geghard village, whose origins date back to ancient times, has been an autonomous municipality since 1986, when it was separated from Goght. The reason for the previous unification of the two villages relies on the intention—during the soviet period—to create a typical state farm (a Kolchoz). At present, the village has 400 inhabitants and 151 buildings. The population is employed in breeding farm and orchard cultivation activities. Still, these occupations are experiencing economic difficulties due to the lack of irrigation water and the scarce profits derived from the meat.

As for the basic supplies to the population, running water pipes are present and working, but there is no sewage system. The village, set along trekking routes, has no signage and lacks a centrality that could serve as a reference point for the population.

References

AA VV (1970) Il complesso monastico di Haghbat, Documenti di Architettura armena. Ares, Milano

Alpago Novello A, Ghalpakhtchian O Kh (1970) Il complesso monastico di Sanahin (10–13 sec), Documenti di Architettura armena. Ares, Milano

Brady Kiesling J (2001) Rediscovering Armenia: an archaeological touristic gazetteer and map set for the historical monuments of Armenia. Tigran Mets, Yerevan

Ghirimezi H (n.d.) Description of Hakhpat and Sanahin, Mèmoires de l'Académie imperial des Sciences de St. Petersbourg, T-Vi, n 6, St. Petersburg (in russian)

Jean de Crimè E (1863) Description des monastères arméniens de Haghpat and Sanahin avec notes et appendice par M. Brosset. St. Petersburg

Sahinian A (1958) Garni/Gegard. Gos. Izd. Isskustvo, Moskva

Sahinian A, Manoukian A, Aslanian AT (1973) G(h)egard. Documenti di Architettura armena. Ares, Milano

UNESCO, world heritage convention, monasteries of Haghpat and Sanahin, documents. https://whc.unesco.org/en/list/777/documents/ . Accessed 14 Jul 2022

UNESCO, world heritage convention, monasteries of Geghard and the upper Azat valley, documents. https://whc.unesco.org/en/list/960/documents/. Accessed 14 Jul 2022

Chapter 5
Toward a People-Based Tourist Approach. Suggestions for the Conservation and Enhancement of the World Heritage Sites and Their Territories

Abstract The inscription of a site on the World Heritage List brings it to the international spotlight, thus opening it to tourism from the "rich" countries. Both governments and populations welcome these processes, due to the economic benefits connected. However, the effects of massive tourism on fragile and sensitive territories can be devastating for Cultural Heritage and for local communities as well. A heritage-based approach is necessary to protect the identity of the places and plan a sustainable development that can bring real benefits to the inhabitants. Only an integrated approach can put the three Monasteries at the center of a virtuous process, based on the assumption that improving the population's living conditions is essential for preserving the Architectural Heritage and developing the tourism offer. This chapter describes the contents of the proposed Master Plan for the Heritage Sites of Haghpat, Sanahin, and Geghard, while highlighting difficulties and possibilities of developing their contexts.

Keywords Master plans for the world heritage sites · Guidelines for the preservation · Tourism development · Monastic complexes · Widespread heritage · Heritage-based approach

5.1 A Master Plan for the World Heritage Sites: General Contents

The three Master Plans for the World Heritage Sites of Haghpat, Sanahin, and Geghard have a similar structure that was shaped by the problems encountered during the study phases of the context.

First, the documents define the Guidelines for the conservation and structural strengthening works of the three monasteries. Indeed, adequate preservation and maintenance of the WHS have a fundamental role in enhancing the places.

The monasteries of Haghpat, Sanahin, and Geghard urgently need studies, investigations, and conservation works to address the current degradation problems that can compromise their future.

The executive projects to preserve them must, therefore, be implemented by the winners of a national or international call for tenders/projects to be issued and managed by the Ministry of Culture (MoC).

The projects should then be implemented based on the guidelines developed in the general plans and necessarily accurate field studies.

The general methodological approach should only solve deterioration problems, avoiding any reconstruction or, in general, any intervention that may distort the integrity and authenticity of the buildings, thus threatening the reading of historical stratifications. A tailor-made maintenance plan must accompany all the conservation projects, in order to identify the necessary periodic controls and ensure the sustainability of the project.

In addition, the Master Plan identifies the most urgent interventions that concern the public spaces around the monasteries to improve the use of these areas by local inhabitants and the level of the tourist experience.

In particular, some interventions are required to improve public space quality. The Master Plan redesigns the facades and fences for buildings facing access roads leading to the Sites, with the aim of utmost improvement of permeability and integration of the villages with the monasteries. Repairing the streets surrounding the WHS and increasing the public lighting network can give greater visibility to the routes and facilitate site access. Parking spaces are arranged at points that do not affect the visual perspective from the site to the surrounding landscape. Removing temporary and abusive structures in the vicinity of the sites can allow more precise reading and perception of the monasteries.

The Master Plans also focus on enhancing the visitor experience in terms of responsible tourism, by involving the population and the widespread Architectural Heritage through creating new museums or improving existing ones. Among these are the Mikoyan Brothers Museum in Sanahin, the Stepan Shahumyan House Museum in Haghpat, and other museums in Goght. In this sense, reusing existing buildings that are culturally attractive or strategically located, can be a valuable support for creating new tourism facilities. By way of example, the former post office building, the Rectory and a vernacular building in Haghpat, the former Café Sanahin, and the Goght Town Hall, could all be pilot projects for tourism purposes. Creating specific itineraries could improve the development of contexts in the vicinity of the monastic complexes. Small interventions in the public space, such as repairing the road surface or installing road signs, could give new quality to the places and encourage an itinerant sale system of local products and souvenirs.

Finally, the Master Plans deal with a considerable part of the villages near the three Monasteries. In particular, it provides for some crucial interventions aimed at increasing hospitality in related areas: the identification of urgent infrastructural interventions to meet basic needs such as water supply and sewerage, the removal of harmful materials (asbestos) from the building roofs, and the preservation of traditional private houses.

All these actions define specific details for the three sites, considering the places' characteristics and specificity as well as their impact on the surrounding area. The investigation also focuses on the relationship with the landscape (the historical villages for Haghpat and Sanahin; the rugged mountains and the natural landscape at Geghard) and their distance to the capital Yerevan, to design different levels of experiences.

Among the three sites, the Geghard Monastery is the first and most visited by international tourists and a traditional destination for domestic tourism, due to its proximity to Yerevan. Therefore, the Master Plan aims to strengthen its role in disseminating Armenian monastic culture.

The village of Goght, located a few kilometers away from the Monastery, has been involved in the process as an ideal place to set up an access point to the Archaeological, Architectural, and Natural Heritage of the area. Haghpat and Sanahin are considered as one system for creating new tourist infrastructure, due to their mutual proximity. Haghpat will welcome tourists by offering a visitor center; while Sanahin will host a didactic workshop offering the opportunity of a learn-by-doing approach to the culture and traditions of Medieval Armenia.

5.2 Guidelines for the Preservation of the Monasteries

Accurate studies and large-scale surveys on the three Monasteries have allowed to develop Guidelines for the preservation and enhancement of these sites. This tool represents a helpful base for subsequent conservation projects. Together with on-site analysis, these studies allowed to define a detailed diagnostic plan for each Monastery and established an initial hypothesis of interventions, which will then be validated based on the results.

The buildings show structural damage of some significance: cracks, detachment, and rotation of the stone ashlars from the underlying masonry in the case of Haghpat (Fig. 5.1); deep cracks in the case of Sanahin (Fig. 5.2) and Geghard (Fig. 5.3).

These structural conditions require an accurate diagnostic project: in-depth crack monitoring, probings and endoscopic tests to define the wall section; georadar and ashlar stone beating to test the detachments.

The integrated methodological approach provides, together with these diagnostic tests, to study all historical sources to understand whether the structural phenomena are progressing.

In all Monasteries, the first review of historical materials shows that the cracking pattern has been stationary for years. The buildings have thus stabilized in their structural damage, but the high seismicity of the Armenian territory suggests the need for urgent, continuous monitoring of current conditions.

Regarding the seismic assessment, a thorough examination of the seismic events that have affected the Armenian territory could provide valuable data.

In particular, the evaluation of qualitative data (induced damage and therefore macroseismic intensity) and, when possible, quantitative information (Peak Ground Acceleration,—PGA, or magnitude), can help reconstruct the general framework. In addition, the evaluation of the strongest earthquake (max PGA) and the consequent

A. The church of Sourb Nshan
B. Large gavit
C. St. Gregory's church
D. The Virgin's chapel
E. Passage and khatchkar of the Saviour
F. Library
G. Hamazasp's building
H. Bell-tower
I. Refectory
L. Tombs of Oukanants
M. Entrances

	RAINWATER SEEPAGE		LOSS AND COLLAPSE OF THE ROOF ELEMENTS
	CRACK, FRACTURE ON HORIZONTAL PLAN		PLANTS WITH ROOTS IN THE WALL (mapped with a line)
	CRACK, FRACTURE ON VERTICAL PLAN		PLANTS ON THE SURFACE (mapped with a line)
	DETACHMENT, ROTATION OF THE STONES		PLANTS ON THE ROOF (definition of the area)
	DISCONTINUITY OF THE MASONRY		DECAY OF THE WALL (Plants, loss of elements, joints erosion both on the top and surface)
	SLIPPING OF THE TILES (definition of the area)		DECAY OF PAINTING (fresco) (erosion, discoloration)
	COLLAPSE OF THE WOODEN ROOF		Note_ The facade surface (both exterior and interior) are covered in some points by a soiling of calcium carbonate, in particular when there is rainwater seepage

Fig. 5.1 Monastery of Haghpat, decay survey of the site (elaboration by authors)

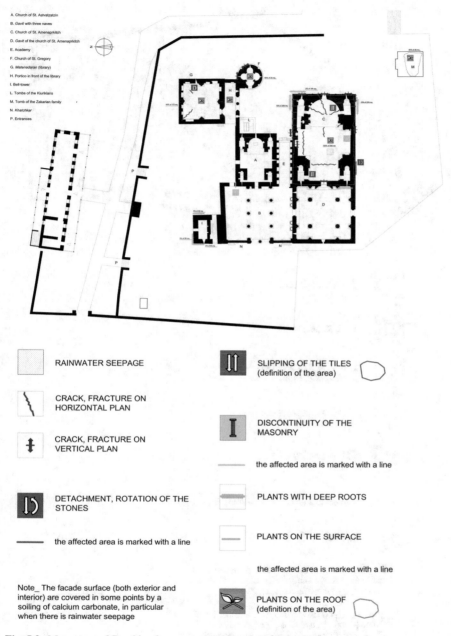

A. Church of St. Astvatzatzin
B. *Gavit* with three naves
C. Church of St. Amenaprkitch
D. *Gavit* of the church of St. Amenaprkitch
E. Academy
F. Church of St. Gregory
G. *Matenadaran* (library)
H. Portico in front of the library
I. Bell-tower
L. Tombs of the Kiurikians
M. Tomb of the Zakarian family
N. *Khatchkar*
P. Entrances

▦ RAINWATER SEEPAGE	⇵	SLIPPING OF THE TILES (definition of the area) ⬡
⸜ CRACK, FRACTURE ON HORIZONTAL PLAN		
↕ CRACK, FRACTURE ON VERTICAL PLAN	Ⅰ	DISCONTINUITY OF THE MASONRY
	——	the affected area is marked with a line
⫧ DETACHMENT, ROTATION OF THE STONES	⬅➡	PLANTS WITH DEEP ROOTS
—— the affected area is marked with a line	—	PLANTS ON THE SURFACE
		the affected area is marked with a line
Note_ The facade surface (both exterior and interior) are covered in some points by a soiling of calcium carbonate, in particular when there is rainwater seepage	🌿	PLANTS ON THE ROOF (definition of the area) ⬡

Fig. 5.2 Monastery of Sanahin, decay survey of the site (elaboration by authors)

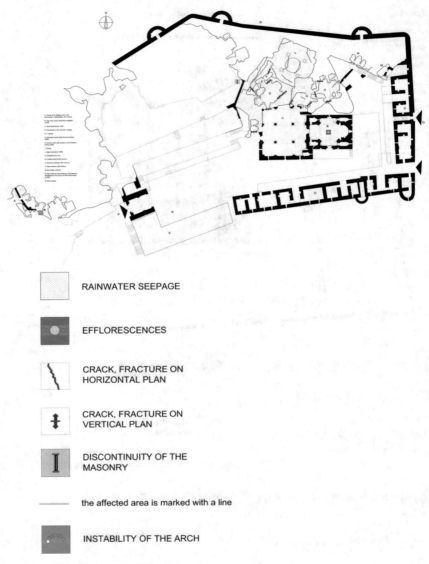

RAINWATER SEEPAGE

EFFLORESCENCES

CRACK, FRACTURE ON
HORIZONTAL PLAN

CRACK, FRACTURE ON
VERTICAL PLAN

DISCONTINUITY OF THE
MASONRY

the affected area is marked with a line

INSTABILITY OF THE ARCH

Note_ Some parts of the facades (both exterior and
interior) are interested by biological colonization in
particular when there is rainwater seepage

_ The facade surface (both exterior and interior) are
covered in some points by a soiling of calcium
carbonate and efflorescences, in particular when
there is rainwater seepage

Fig. 5.3 Monastery of Geghard, decay survey of the site (elaboration by authors)

behavior of monuments (collapses, types of damages, severity) must be carried out by considering the improvement of structural properties thanks to the masonry grouting.

The disintegration of mortars in rubble masonry is, in fact, a severe problem for the Monastery's structure, due to water infiltrations and the penetration of soluble salts. During the development of the detailed project, injections of hydraulic lime mortar into the masonry to restore its strength and sealing existing cracks so as to prevent rainwater penetration are fundamental interventions.

Analyzing existing mortars in terms of chemical and physical characteristics can be essential to define the composition of materials and establish compatibility with new mixtures for the injections.

The surfaces of the three religious complexes are, in all cases, affected by black compact deposits and white formations of soluble salts (Figs. 5.4, 5.5 and 5.6).

The conservation project must therefore provide cleaning works by using non-invasive techniques. Only after some specific tests will it be possible to identify the most effective methods to remove the concretions of calcium carbonate, deposits, and soluble salts.

Some specific decay and structural problems have been identified in the three Monasteries, leading to different indications for the interventions.

Fig. 5.4 Monastery of Haghpat, the bell tower with an evident out of plumb (Photo M. Giambruno 2015)

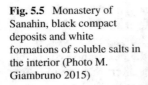

Fig. 5.5 Monastery of
Sanahin, black compact
deposits and white
formations of soluble salts in
the interior (Photo M.
Giambruno 2015)

In **Haghpat**, the severity of the structural framework's conditions requires preventing cracks opening by using steel ties. In the church of St. Nshan and other parts of the Monastery affected by large cracks (e.g., more than 20 mm), confinement reinforcement rings are needed, with stitching theme by insertion of inclined (steel or carbon fiber) bars in holes drilled in the masonry.

For preserving the roofs, particular importance has to be devoted to studying a waterproofing system to prevent infiltration of meteoric waters inside the buildings while allowing moisture evaporation. Other punctual interventions are requested to repair the roofs: removing the weed vegetation that could cause disruptions, replacing cracked tiles, and grouting the missing parts with lime mortar. Moreover, the water drainage system needs redesigning with correct inclinations to facilitate the outflow and avoid direct drainage on the façade (Figs. 5.7 and 5.8).

The overall external surfaces and the remains of the frescoes and mortars need careful cleaning to remove compact deposits (calcium carbonate) and soluble salts, mainly produced by rainwater infiltrations.

Some preliminary cleaning tests are essential to assess the effectiveness of the intervention and the application time of materials. Wraps of sepiolite and deionized water, compresses of sepiolite with water and ammonium carbonate, or laser cleaning are the best products.

Fig. 5.6 Monastery of
Geghard, formations of
soluble salts in the upper part
(Photo M. Giambruno 2015)

Cleaning the remains of frescoes requires accuracy. Before the interventions, a careful study of the constitutive materials (e.g., the types of pigments), the support, and the previous restoration works are essential as knowledge support. After acquiring all data, a cortical consolidation can be helpful for substrate re-adhesion.

Other problems have been detected during the survey phases. For example, the walls of the Monastery have been reconstructed in different places, and the ridges are almost entirely missing. The current entrance has been realized in recent years through partial demolition of the walls, and other parts have been rebuilt. These repeated interventions have generated some critical points in the connections.

It is urgent to consolidate the entire walls, especially at the points where the stones are no longer bound.

Moreover, a soft capping of the masonry walls with hydraulic lime mortar is needed to protect the wall tops. In the tower on the right of the Monastery entrance, the two barrel vaults that gave access to an underground tunnel built to favor the escape are still visible. It is unthinkable to reopen this route, which essentially collapsed. However, it would be interesting to study solutions to increase the visibility of such access (Figs. 5.9 and 5.10).

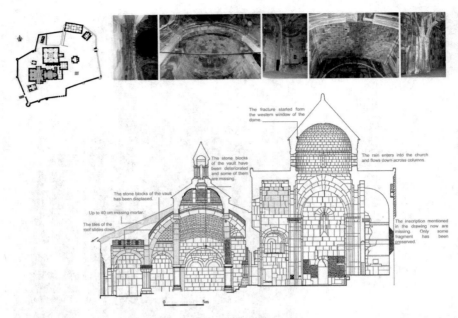

Fig. 5.7 Monastery of Haghpat, section to the north of large gavit and church St. Nshan (elaboration by authors)

In Sanahin, the consolidation of the architrave is urgent. The insertion of inclined steel bars in holes drilled in the stone and subsequently filling with hydraulic fiber-reinforced mortar could prevent flexural failure of the lintel. In the church of St. Amenaprkich, already compromised, interventions with steel ties and confinement reinforcement rings can prevent further cracks (Figs. 5.11, 5.12 and 5.13). Horizontal steel ties can connect weakly linked parallel walls and absorb the pushing forces at the arch reins. Circular rings made of steel plates can absorb the horizontal forces at the dome's base.

In the Sanahin complex, pavements are a crucial issue. An archaeological survey with deep inspections is essential before proceeding with any conservation works. Once the data collected, it will be possible to fill the missing parts with stone labs or, preferably, by realizing a false beaten-earth floor with a mix of lime mortar and earth. This substrate ensures maximum compatibility with the existing pavement and, simultaneously, allow its use and facilitate the realization of horizontal paving comfortable for walking on.

Any fractures in the original stones can be fixed by removing the parts, stitching the fractures with fiberglass pins, and repositioning the stones.

The restoration works (also the most recent ones) already implemented in **Geghard** have improved the overall situation of the complex. Some problems are still present, particularly cracks found in the underground zones included in a past project. In this case also, it is urgent to intervene in the short term. Filling the cracks

Fig. 5.8 Monastery of Haghpat, north facade of the large gavit and church St. Nshan (elaboration by authors)

to avoid rainwater infiltrations is crucial to prevent further decay. Moreover, monitoring the sealings made over time is essential for controlling the results. Before starting any conservation works, solving the water penetration problem is necessary. Only compatible materials must be used to fill the cracks, such as lime mortars, and plastic or synthetic products that can trigger further decays must be avoided. Mortar must have an elastic modulus capable of bearing the inevitable deformations. The Geghard Monastery shows a critical situation from the point of view of safety. Highly fractured rocks around the complex and along the access way make the place unsafe for tourists, requiring urgent interventions, both passive and active (using, for example, nets against rockfalls or nets fixing the rocks). However, a preliminary geomechanical structural survey is essential for planning effective and non-invasive interventions. The knowledge process must identify the most dangerous rock zones (i.e., the zones with the most significant number of fractures or the widest fractures) where to focus the primary and most important interventions. Moreover, it will be necessary to set up a continuous monitoring system of rock fractures using crack

A. The church of Sourb Nshan

B. Large gavit

C. St. Gregory's church

D. The Virgin's chapel

E. Passage and khatchkar of the Saviour

F. Library

G. Hamazasp's building

H. Bell-tower

I. Refectory

L. Tombs of Oukanants

M. Entrances

Fig. 5.9 Monastery of Haghpat, conservation project (elaboration by authors)

meters. This system can detect incremental movements (not only due to seasonal temperature changes) and, in case, allow to fix an alarm threshold referred to a high risk of rockfall (Figs. 5.14, 5.15 and 5.16).

5.3 Developing the Villages. Improving the Living Conditions of the Population to Enhance the Tourism Industry

International documents on cultural tourism require tourism to benefit the local communities. As early as in 1999, UNESCO adopted the International Cultural Tourism Charter, clarifying the aims and objectives of cultural tourism. It established that Cultural Heritage must be made accessible not only to tourists but also

GUIDELINES FOR THE
CONSERVATION PROJECT
STEPS

PRELIMINARY DIAGNOSTIC TESTS

PRELIMINARY WORKS:
- Removals
- Dry cleaning

CONSOLIDATION WORKS

CLEANING WORKS

PROTECTION WORKS OF THE SURFACES

PUNCTUAL INTERVENTIONS

DIAGNOSTIC TESTS

INTERVENTION ON THE ROOF
- Removal of tiles
- Manual removal of soiling and disintegrated mortar
- Removal of vegetation. For deep roots injections of proper products (metossitriazine)
- Cleaning of the elements
- Laying a layer of hydraulic lime mortar with pumice stone or tuff (3 cm)
  (taking care of the water runoff)
- Replacement of the missing parts with similar elements
- Eventual elastic joint in relation to the shape and size of tiles
The slope and the water drain system must be designed avoiding the sliding of the water on the facade

INJECTIONS OF HYDRAULIC LIME MORTAR
(of general wall section)

ENSURE THE SAFETY OF THE WOODEN ROOF, REPLACING DETERIORATED PARTS
- Recovery of the wooden elements with appropriate treatments

MONITORING THE CRACKS.
IF STABLE, REPAIR FILLING THEM WITH HYDRAULIC FIBRE-REINFORCED MORTAR
IF WIDE, INSERT ALSO DIAGONAL STEEL BARS TO SEW THE CRACK

GENERAL CLEANING OF THE SURFACES WITH LOW PRESSURE WATER

CLEANING OF CALCIUM CARBONATE SOILING WITH
- COMPRESSES OF SEPIOLITE
- COMPRESSES OF AMMONIUM CARBONATE AND SEPIOLITE
- CLEANING WITH LASER
(After appropriate cleaning tests)

REMOVAL OF THE ROOTS WITH INJECTIONS OF PROPER PRODUCT (METOSSITRIAZINA)
- Manual uprooting after the action of the product

INTERVENTIONS ON THE WALL
- Removal of vegetation (for deep roots injections of proper product - metossitriazina)
- Grouting of joints with hydraulic lime mortar with pumice stone or tuff
- Definition of the wall tops by capping (with proper slope)

CONSERVATION OF FRESCO

INTERVENTION ON THE FLOOR
- Archaeological survey of all stones (1:10; 1:20)
- Conservation project of the existing stones;
- Creation of a new coplanar plane
- Integration project of the missing parts

Note:
MORTAR
The type of mortar for conservation interventions must be selected in relation to the elastic modulus and coefficient of deformability with regard to the existing materials (lime mortar with ground pumice or tuffs and fiber to improve the elastic modulus)

Note:
Supporting the structural monitoring it could be useful a detailed analysis of the historical pictures.
From a preliminary examination the cracks were already existing and apparently they didn't changed significantly

Fig. 5.10 Monastery of Haghpat, conservation project, legend of the main interventions (elaboration by authors)

to the local community, and tourism must be respectful of host communities and their "living cultures". In its fourth principle, the Charter clarifies how to involve the local communities in the planning of tourism development (ICOMOS 1993, 2002). A more recent document from the European Commission, Directorate General for Education, Youth, Sport and Culture, states that "Sustainable cultural tourism is the integrated management of cultural heritage and tourism activities in conjunction with the local community, creating social, environmental and economic benefits for all stakeholders to achieve tangible and intangible cultural heritage conservation and sustainable tourism development" (EU 2019).

This perspective is, unfortunately, far from the situation of the inhabitants living near the three WHS monasteries. As mentioned in the previous chapter, they often live in subsistence conditions, benefiting very little from the tourist flows that quickly visit these sites to then go back and consequently spend their money in the capital. For these reasons, the project aims to act on a double line. From one side, it intends to

Fig. 5.11 Monastery of Sanahin, orthophoto of the southern façade of St. Amenaprkich (elaboration by authors)

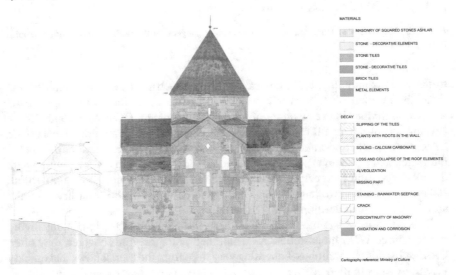

Fig. 5.12 Monastery of Sanahin, material and decay survey of the southern façade of St. Amenaprkich (elaboration by authors)

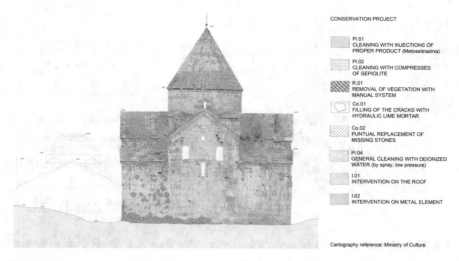

Fig. 5.13 Monastery of Sanahin, conservation project of the southern façade of St. Amenaprkich (elaboration by authors)

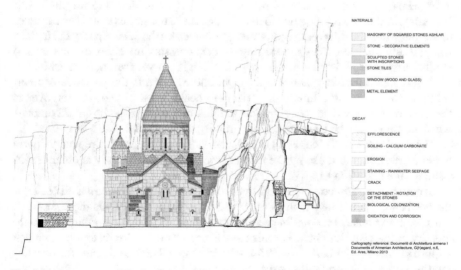

Fig. 5.14 Monastery of Geghard, material and decay survey of the main facade (elaboration by authors)

propose a series of valorization interventions for community-based tourism, which we will discuss in the next section.

On the other side, it proposes direct interventions in the villages based on the assumption that improving the population's living conditions is essential for preserving the Architectural Heritage and developing the tourism offer.

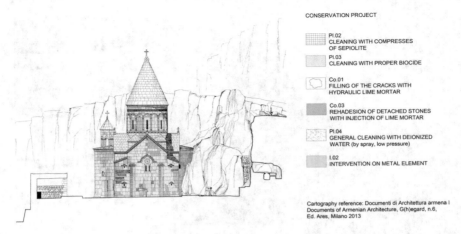

Fig. 5.15 Monastery of Geghard, conservation project of the main facade (elaboration by authors)

In Haghpat, the construction of a network for the distribution of running water is of primary importance. Its absence still forces the inhabitants to use the public fountains. Part of the funding intended to restore the Monasteries could be channeled to this purpose. The intervention was already planned by the Municipality of Haghpat and partially financed. They are looking for the necessary matching of the funds. A similar priority regards the sewage system, which is not yet present in Sanahin.

A significant problem plaguing Haghpat and Sanahin is the presence of many asbestos roofs. The harmful effects on people's health are well-known. Moreover, in these conditions, it would be impossible to think of a widespread hospitality model, which could instead be an excellent opportunity for these places.

The conditions of the public lighting and roads, and the need to repave them for better accessibility and use by the residents including in winter, is another crucial issue for the two villages.

In both places, the economic conditions of the resident population are difficult. On the one hand, it is not easy to think of private financial resources. On the other hand, it is urgent to implement coordinated and consistent works that do not transform the identity of the historic center. For this reason, one or more pilot restoration projects on historic residential buildings could help develop general guidelines for carrying out the works, including self-construction.

In Sanahin, the reopening of the funicular that connects the village to the town of Alaverdi is a matter of central importance. In fact, Alaverdi is currently a separate administrative entity.

This intervention would allow residents to connect with the city even in the winter, when the roads are often impracticable. At the same time, tourists could rely on an ecological means of transport to reach the Monastery while also admiring the valley and the Industrial Heritage of the city.

GUIDELINES FOR THE
CONSERVATION PROJECT
STEPS

PRELIMINARY DIAGNOSTIC TESTS

PRELIMINARY WORKS:
- Removals
- Dry cleaning

CONSOLIDATION WORKS

CLEANING WORKS

PROTECTION WORKS OF THE SURFACES

PUNCTUAL INTERVENTIONS

DIAGNOSTIC TESTS

INTERVENTION ON THE ROOF
- Analysis and redesign of the pitches and of the water drain system to favor water run off and avoid its draining on the facade
- removal of old and deteriorated pointing realized with plastic materials
- Removal of biological colonization

INJECTIONS OF HYDRAULIC LIME MORTAR

REMOVAL OF EFFLORESCENCES BY BRUSHING
(eventual compresses for the extraction of salts)

MONITORING THE CRACKS.
IF STABLE, REPAIR FILLING THEM WITH HYDRAULIC FIBRE-REINFORCED MORTAR
IF WIDE, INSERT ALSO DIAGONAL STEEL BARS TO SEW THE CRACK

GENERAL CLEANING OF THE SURFACES WITH LOW PRESSURE WATER

PROVISIONAL STRUCTURES TO ENSURE THE STABILITY OF THE ARCH

REMOVAL OF BIOLOGICAL COLONIZATION BY APPLICATION (BY SPRAY OR BRUSH) OF PROPER BIOCIDAL PRODUCTS

CLEANING OF CALCIUM CARBONATE SOILING WITH
- COMPRESSES OF SEPIOLITE
- COMPRESSES OF AMMONIUM CARBONATE AND SEPIOLITE
- CLEANING WITH LASER
(After appropriate cleaning tests)

Note:
MORTAR
The type of mortar for conservation interventions must be selected in relation to the elastic modulus and coefficient of deformability with regard to the existing materials (lime mortar with ground pumice or tuffs and fiber to improve the elastic modulus)

Note:
Supporting the structural monitoring it could be useful a detailed analysis of the historical pictures.
From a preliminary examination the cracks were already existing and apparently they didn't changed significantly

Note:
General check of the efficiency of water barrier system realized _____

Fig. 5.16 Monastery of Geghard, conservation project of the site (elaboration by authors)

Fig. 5.17 Village of Goght, survey of the critical points of the main square (Photo by authors, 2015)

The village of Goght is lower than the main road leading to the Monastery of Geghard. Goght is not currently involved in the tourism circuit, and its inhabitants benefit only partially from the economic flow generated by tourists who visit the Monastery. The village's central square, just a few kilometers from the Monastery, could become a new polarity to involve the center in the tourism circuit and allow inhabitants to share its benefits (Fig. 5.17).

Entirely redesigning the square with new paving and lighting could allow the community to enjoy a unique meeting place and a recognition point for inhabitants. This intervention could generate a virtuous enhancement of the entire village, improving social cohesion. At the same time, this new polarity could host a museum or a cultural center which would increase the flow of visitors, making better use of available resources and decreasing tourism pressure on the Monastery. The important modernist building in the town square, now only partially used as the Town Hall, could accommodate the new functions (Fig. 5.18). After adequate rehabilitation as a museum/visitor center, the complex could introduce visitors to the Armenian Cultural Heritage, especially that related to the beginning of local civilization. The Armenian Heritage linked to prehistory is not very well known. Several archaeological missions conducted in the last twenty years, often with the support of American universities, have found very interesting remains able to change our vision of early history. The museum could show the public the wealth of the ancient Archaeological Heritage discovered in Armenia, starting from information about the petroglyphs scattered along the upper slopes of the Azat river. The second floor might host a civic center

Fig. 5.18 Village of Goght, the modernist building in the Town square, now partially used as Town Hall (Photo by authors, 2015)

with a meeting hall, chess tables, internet spots, a youth center, etc. The Geghard village, due to its vicinity with the Monastery, may also be included in the tourism circuit. Paving the main roads, improving the public lighting system, and realizing a sewage system are necessary. The surrounding landscape made up of fruit orchards suggests the possibility of developing farm holiday activities.

The agricultural vocation of Goght and Geghard may be improved and favored by realizing a public market on the main road for selling local fruit and flowers.

5.4 The Monastic Complexes: Enhancement and Fruition

The three monastic complexes need coordinated interventions to improve their fruition and distribute tourist flows to the various sites' points. These would also be useful cautions to avoid saturating the monasteries' carrying capacity and preserve their authenticity. The project aims at developing four topics:

- the fences that close the complexes, in particular for Haghpat and Sanahin;
- signage and information boards;
- waste management;
- lighting of the architecture in the complexes.

The existing signage is inefficient and not always visible. By way of example, redesigning signs as small elements to be placed on the ground could reduce their impact. More specific brochures are also needed.

The QR code technology, cheap and easy to spread, could guide visits and provide explanatory information. A small metal or stone plate embedded in the pavement could be less invasive and provide more information than traditional signage.

A specific project should address waste management, especially where large numbers of tourists are expected.

The long rows of garbage bins that welcome tourists at the entrance of the Sanahin complex are not a good model.

The lighting design to enhance the architecture of sites has become essential over the years and the subject of specific studies and projects. This aspect is crucial for the three monasteries. In particular, for Haghpat and Sanahin, the lighting of architecture could reinforce the visual polarity, giving new centrality to the two surrounding settlements.

In addition to these general indications concerning the three complexes, the Master Plan envisages specific interventions for each of them.

In **Haghpat**, a lawn crossed by narrow dirt paths surrounds the monumental complex. This element, pleasant and appropriate to the site, should be preserved. New low fruit trees could embellish the two terraces adjacent to the monastery walls. In the monastic enclosure, the small thuja tree that lines the access to the refectory and the staircase connecting the two levels does not seem appropriate to the context. Avoiding planting trees within the enclosure would be more suitable.

The project must also focus on the iron fences (railing separating the refectory from the street, access gate) by designing actions coordinated with all other interventions (Fig. 5.19).

A wall fence borders the complex of **Sanahin**, surrounded by a lawn area that requires better maintenance and homogeneity with the rest. The souvenirs kiosk inside of the monastic complex walls needs redesigning and a better position, maybe immediately outside the borders. The trees that harmoniously surround the complex should be preserved and maintained, in addition to planting new species.

The fences at the site entrance and some railings in steep position required redesigning to harmonize with the site (Fig. 5.20).

The site's entrance a prominent metallic structure corroded by atmospheric agents is found that was used for previous restoration works. Removing it could improve both access to the site and the visiting experience.

In the modern cemetery on the back of the complex the remains are found of three old little churches dating from the tenth to the thirteenth centuries. One is in a state of ruin, while the others are in discrete conditions. Preservation works and well-designed signs could be added to the visit itinerary.

The ancient library of the monastic complex is exciting both for its architecture and the materials conserved therein. The guardian reports that the library previously had clay pots (*karas*) buried in the ground.

After archaeological excavations, no information remained in the Monastery, and only the findings of minor importance are displayed there. The wooden door is recent,

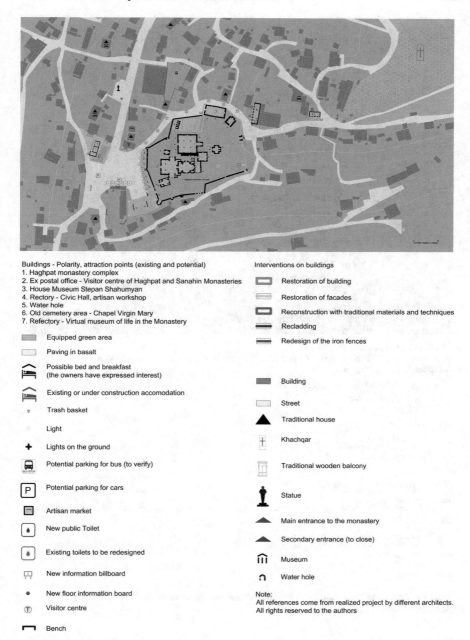

Buildings - Polarity, attraction points (existing and potential)
1. Haghpat monastery complex
2. Ex postal office - Visitor centre of Haghpat and Sanahin Monasteries
3. House Museum Stepan Shahumyan
4. Rectory - Civic Hall, artisan workshop
5. Water hole
6. Old cemetery area - Chapel Virgin Mary
7. Refectory - Virtual museum of life in the Monastery

- Equipped green area
- Paving in basalt
- Possible bed and breakfast (the owners have expressed interest)
- Existing or under construction accomodation
- Trash basket
- Light
- Lights on the ground
- Potential parking for bus (to verify)
- Potential parking for cars
- Artisan market
- New public Toilet
- Existing toilets to be redesigned
- New information billboard
- New floor information board
- Visitor centre
- Bench

Interventions on buildings

- Restoration of building
- Restoration of facades
- Reconstruction with traditional materials and techniques
- Recladding
- Redesign of the iron fences

- Building
- Street
- Traditional house
- Khachqar
- Traditional wooden balcony
- Statue
- Main entrance to the monastery
- Secondary entrance (to close)
- Museum
- Water hole

Note:
All references come from realized project by different architects.
All rights reserved to the authors

Fig. 5.19 Monastery of Haghpat, guidelines for the enhancement of the visiting experience (elaboration by authors)

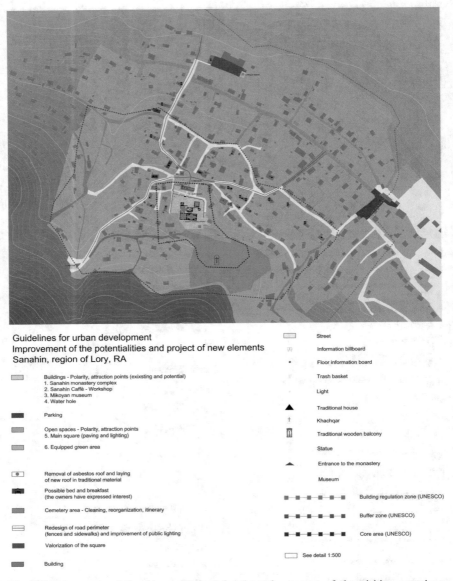

Guidelines for urban development
Improvement of the potentialities and project of new elements
Sanahin, region of Lory, RA

Buildings - Polarity, attraction points (exixsting and potential)
1. Sanahin monastery complex
2. Sanahin Caffè - Workshop
3. Mikoyan museum
4. Water hole

Parking

Open spaces - Polarity, attraction points
5. Main square (paving and lighting)

6. Equipped green area

Removal of asbestos roof and laying
of new roof in traditional material

Possible bed and breakfast
(the owners have expressed interest)

Cemetery area - Cleaning, reorganization, itinerary

Redesign of road perimeter
(fences and sidewalks) and improvement of public lighting

Valorization of the square

Building

Street

Information billboard

Floor information board

Trash basket

Light

Traditional house

Khachqar

Traditional wooden balcony

Statue

Entrance to the monastery

Museum

Building regulation zone (UNESCO)

Buffer zone (UNESCO)

Core area (UNESCO)

See detail 1:500

Fig. 5.20 Monastery of Sanahin, guidelines for the enhancement of the visiting experience (elaboration by authors)

while the original piece is currently on display at the National Museum of History. The actual destination of the small museum as a *lapidarium* is coherent with the context, so the project envisages to consolidate it.

Some specific interventions also concern **Geghard**, based on its architectural features and the trend of visitors' inflow (Fig. 5.21). The site is crowded with tourists even in winter, due to its proximity to Yerevan, its traditions, architectural quality, and natural environment. The space between the churches, *gavit,* and the service buildings (offices, monks' dwellings, small bookshop, etc.) is enjoyable.

On the left, three terraces interrupt the slope. Specific actions on this point could further enhance the environment. The first terrace is stone-paved and gives access to a room partially made from rock and not currently in use.

The building between the two terraces, which is partially underground, could become a singing school to enhance an activity already present in the Monastery but not yet correctly organized. The acoustics inside the upper church is quite particular and enhances the singers' voices, providing listeners with a moving and unforgettable experience. The second terrace is stone-paved and could be improved to be used as a resting area for visitors.

The monks use the third terrace as an orchard with hives for beekeeping. Its reorganization and at least partial opening to the public could document some of the monks' activities in their everyday lives.

5.5 Preserving the Widespread Heritage to Improve the Touristic Experiences

Tourism development in the territories around the Monasteries requires new facilities to improve knowledge of the place, diversify and make the experience more enjoyable. A better quality of the offer could encourage visitors to stay longer in the villages and their surroundings.

In addition to the Monastery and the square, **Haghpat** also offers other exciting places to visit: the post office building (Fig. 5.22), a traditional architecture placed below the plaza; the Stepan Shahumyan house-museum; the old water hole and the chapel of the Virgin Mary just outside the village. The connection path requires good paving and signs. If adequately rehabilitated, the square in front of the entrance gate to the complex can become a new polarity for the entire local community. The inhabitants can also benefit from the nearby Town Hall Square. Removing and replacing the kiosks in the southern edge could enhance the experience, also enriched by the beautiful view of the valley.

The monastic complex and its surroundings could also host new reception points for tourists, simply recovering existing buildings and avoiding the construction of new buildings. Hospitality could be managed by the resident population. The refectory may host a small exhibition on the history of the Monastery, its architecture, and its features. The museum's project should provide for display elements on the ground

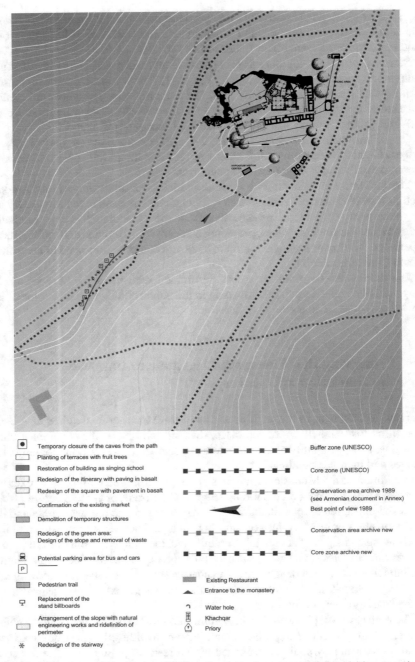

Temporary closure of the caves from the path

Planting of terraces with fruit trees

Restoration of building as singing school

Redesign of the itinerary with paving in basalt

Redesign of the square with pavement in basalt

Confirmation of the existing market

Demolition of temporary structures

Redesign of the green area:
Design of the slope and removal of waste

Potential parking area for bus and cars

Pedestrian trail

Replacement of the
stand billboards

Arrangement of the slope with natural
engineering works and redefinition of
perimeter

Redesign of the stairway

Buffer zone (UNESCO)

Core zone (UNESCO)

Conservation area archive 1989
(see Armenian document in Annex)

Best point of view 1989

Conservation area archive new

Core zone archive new

Existing Restaurant

Entrance to the monastery

Water hole

Khachqar

Priory

Fig. 5.21 Monastery of Geghard, guidelines for the enhancement of the visiting experience (elaboration by authors)

Fig. 5.22 Haghpat village, the post office building (Photo by authors, 2015)

or a virtual museum, so as not to interfere with the architecture of the building. This old building, once used as a post office, can become a visitor center. It could also accommodate a bookshop. The access road, which also serves some residential buildings, needs paving. The abandoned building, located on the north of the square, can host a small café offering local products tasting.

The Stepan Shahumian House Museum (Figs. 5.23 and 5.24), now closed, should be rearranged to offer visitors an idea of Armenian life under the communist regime. The room on the ground floor of the Rectory (now partially restored), with new pavements and a heating system, could host courses for people involved in tourism and small laboratories to manufacture souvenirs. A problem that plagues the square of the Monastery is the parking. In some periods, the spaces are so saturated with cars that their use becomes difficult and visitors' perception of the complex is distorted.

Here, car parking should be regulated and only allowed to people with reduced mobility or emergency vehicles. Parking large buses should be prohibited. The place should be only partially paved, to leave a portion of greenery. The slight slope can be exploited to add small low walls as seating places.

A small number of large—and micro—buses could leave the tourists in the square and then move to near the Town Hall. For more significant numbers of vehicles, one proposal can be the expansion of the banked area behind the Town Hall; a second is to provide parking for visitors going to Haghpat and Sanahin at Alaverdi, in a place to be selected with the Municipality.

Fig. 5.23 Haghpat village, Stepan Shahumian House Museum (Photo by authors, 2015)

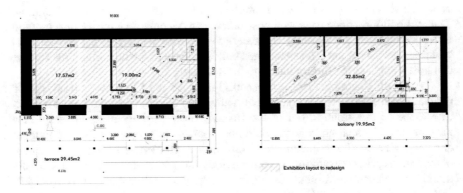

Fig. 5.24 Haghpat village, Stepan Shahumian House Museum, plan of the ground and first floor (elaboration by authors)

Along the road on the left of the monastery wall, an open-air market with removable structures for selling souvenirs, local products, etc., can find place.

Even the **Sanahin** Monastery, just like Haghpat, could be part of a system including other Cultural Heritage, to encourage the involvement of the entire village in tourism activities.

The proposed itinerary touches both the Mikoyan House Museum and the old water hole. The museum has exciting materials, but the exhibition is still obsolete and requires to be entirely redesigned.

The museum should be made interactive by providing flight simulators and other devices for active involvement of visitors in the experience. Furthermore, displaying and selling a good selection of thematic souvenirs on websites can enrich visitor's experience. The renovation of the road's surface that leads to the museum is essential, considering that this road also gives access to a school.

The water hole is located in a small lovely valley, and a trail crossing the village could connect it with the Monastery. A picnic area and a children's playground could be created near the water hole.

The building near the Monastery could host facilities for welcoming visitors.

The "Sanahin café" (Fig. 5.25) has a pleasant architectural structure, with a wide wooden balcony and a long porch looking toward the Monastery. The café is organized into two levels, offering wide spaces for the necessary functions in support of visitors. In addition to public toilets (Fig. 5.26), the café can also accommodate rooms for workshops on the creation of Armenian prayer scrolls, thus recalling the ancient function of the Monastery.

Fig. 5.25 Sanahin, surrounding of the Monastery, the Sanahin Café. The building can be rehabilitated to host tourist facilities (Photo by authors, 2015)

Fig. 5.26 Sanahin, surrounding of the Monastery, the public toilets (Photo by authors, 2015)

Laboratories and workshops for realizing copies of famous manuscripts with natural pigments and traditional techniques to be then sold as souvenirs can complete the tourism offer.

The first floor can host the exhibition and local products tasting (resting point).

Laboratories spaces can attract also school groups from the nearby towns and villages. The general state of the building is quite precarious and requires a conservation project for the building to host new functions. The roofing, made of asbestos, is partially ruined; the porch shows signs of foundation sinking, and the stability of the whole first floor and the wooden balcony need accurate verification.

The quality of the green space in front of the building and the monastery access needs improvement. This area may host seats for the rest of the visitors. These activities should be managed by the local community, after suitable training which could also concern the establishment of scroll/painting laboratories.

The project must pay special attention to redesigning the square at the site's entrance, which now is very narrow. This focus can be the opportunity to enhance access to the monastic complex.

In Sanahin, too, car parking is a problem, due to the size of the entrance area which can only host a very reduced number of cars. A more extensive car and bus park can be realized on the edges of the urban area, at a small distance from the Monastery (10 min' walk). This solution would allow tourists to also visit the village and discover its historical architecture, and solve one of the problems plaguing access to the Monastery.

During summertime, many souvenirs sellers come from the capital and place their stalls along the staircase. Although this little market does not offer local products, it is picturesque; however, the conditions of sellers are inadequate, and the provisional canopies built to protect them from sun and rain are of poor architectural quality. Furthermore, their stalls reduce the space available and obstruct the passage of big groups.

A part of these commercial activities could be moved to the former Sanahin café buildings and a part along the road that connects the parking to the monastic complex. Along this path, the houses with their green areas may host the sale/tasting of local products. Targeted training and Small Medium Enterprises incubators may help inhabitants to start these little business activities.

The restoration of the "soviet" building facing the entrance gate can be a further upgrade of the entrance square. The building, simple but architecturally exciting, already hosts a souvenir shop. Relocating the kiosks in front of the wall around the Sanahin café in this structure might allow better use of the spaces. Moreover, restoring the soviet building while preserving its peculiarities could enhance the global quality of the site.

Finally, the rehabilitation and redesign of the facade and courtyard of the building facing the entrance stairs could complete the interventions. The building now hosts some demonstrative activities to realize local products such as the typical *lavash* bread, liqueurs, etc. These functions should be incentivized, but this requires a comprehensive redesign.

As in the two previous cases, in **Geghard** the problem of the car parking area being too close to the complex's walls is the first critical issue. The parking area should be moved downhill along the road, and access should only be allowed to people with limited mobility.

Locating the car parking area at a distance from the Monastery entails two additional advantages. On the one hand, it leads visitors to walk on the ancient trail along the slope of the mountain, flanking some caves and rooms carved in the rock. These paths need rehabilitation to provide hikers a more pleasant arrival. On the other hand, this proposal would leave more space available for the spontaneous market of typical products, which is Geghard's Intangible Heritage. This market should be maintained in its current position and preserved in its spontaneity. The item to improve is the condition of sellers who currently have to carry their products back and forth every day.

The two unused rooms (former toilets) located just behind the sellers could be rehabilitated so as to be heated in winter and host a small warehouse for the shops as well as a small toilet.

The Geghard Monastery represents the first stop for many tourism circuits in Armenia. For this reason, the site could be the right point for a well-organized visitor center, not only to illustrate the history and potential of the area but also to introduce the history of Armenian monasteries. The new visitor center must not impact the Monastery and the underground churches. A new building can be placed on the access road, covered by earth and grass, and reconstructing the slope of the mountain after the investigation of the geological situation of the area.

5.6   Some Final Considerations

The work presented in this book was commissioned by the World Bank in 2015 as part of the more general Armenia—Local Economy and Infrastructure Development Project. The project primarily aimed at preserving and enhancing the monastic complexes of Haghpat, Sanahin, and Geghard. Haghpat and Sanahin were listed as World Heritage Sites in 1998 and Geghard in 2000.

The monasteries of Haghpat and Sanahin are placed in the North of Armenia, not far from the border with Georgia, while Geghard is the nearest to the capital Yerevan. These geographical locations give rise to different tourist flows. Visitors are more numerous in Geghard, although the UNESCO brand includes all of the three villages in its travel itineraries. The surrounding areas do not benefit economically from tourism. Tourists only visit these places for a few hours, without entering into contact with the local population, and usually stay in the grand hotels of the capital.

Starting from these reflections, the proposal worked on two main objectives. Firstly, understanding the conservation state of the three complexes and defining guidelines for their preservation. Secondly, creating of itineraries around the three monasteries. These circuits could enhance the Tangible and Intangible Heritage which is almost forgotten compared to the three monuments, and improve the economic conditions of the rural communities in the surrounding villages. The construction of primary infrastructural works currently absents, the creation of new community services, and training in tourism and handicrafts could increase these possibilities.

The project aims to place the monasteries of Haghpat, Sanahin, and Geghard, adequately preserved, at the center of a virtuous circle. The availability of quality services could guarantee tourists full enjoyment of their visit. At the same time, local inhabitants could improve their living conditions thanks to the opportunities of a well-structured and international tourism circuit.

Promoting Armenian culture and traditions can enrich the tourist's experience and create further opportunities to involve the local population in the economic benefits generated by tourism.

The inhabitants interviewed during the project showed great interest in widespread hospitality activities and in catering services with local products.

The project intended to enhance the specific features of the villages surrounding the three monasteries, and ensure the conservation of their historical and architectural characteristics. For this reason, the villages represent the core of the proposal for developing sustainable cultural tourism. This place-based approach could limit the construction of "globalized" accommodation facilities of poor quality and attract tourists interested in experiencing the places.

The project included interventions for a broadly estimated amount of 17 million dollars. The overall funding included works described in this volume and all the supporting activities, such as adapting the road network and creating the itineraries. The proposal focused on the three UNESCO sites and the Cultural Heritage of the vast territory surrounding them.

The work implemented in the consultancy activity exceeded by far the specific requests by the World Bank, specifically addressed to the monasteries and their close places.

Extending the reflections to the surrounding villages comes from previous experiences in "emerging countries". The deep conviction is that cultural tourism cannot develop without, or worse, to the detriment of local communities and their resources. The process must start with improving the inhabitants' living conditions, and then spread to a territorial scale.

As already mentioned, this study had no prescriptive character. The aim was to provide the Armenian government with a comprehensive guideline tool based on current international knowledge on the conservation of Architectural Heritage and the development of cultural tourism. The aim of the guidelines was to support the government in the case it decided to channel part of the funding granted by the World Bank to these issues.

However, these studies are often perceived as interference by academic "foreigners" who do not know the country's actual needs.

It is undeniable that there Armenia has many other priorities in, given the young age of this small state and the economic crisis due to the separation and fall of the Soviet Union.

Moreover, as often happens in these circumstances, tourism is seen as a way to increase GDP without thinking of or carefully considering its possible effects on a fragile territory. The environment has been preserved precisely because of the lack of economic resources. Furthermore, "monuments", even more than Cultural Heritage, are only considered as a way to achieve the goal quickly.

Indeed, Tourism and Cultural Heritage are an apparently winning combination for developing a territory (UNEP 2005).

In emerging countries, the legacy from the past is more fragile and therefore more vulnerable to tourism "development". So, which can be the outcomes of these processes? (Musso 2015).

Mass tourism is the first result of a country opening up or investing in this field. The results are well known, not only in emerging countries, and the effects are detrimental to preserving the Cultural Heritage and the natural environment. The need to accommodate large numbers of people leads to building from scratch facilities and infrastructure, often without caring of the existing contexts (UNWTO 2013).

The resident population has little or no benefit from tourism growth, as at most they only obtain low-income jobs within the tourism facilities; in addition, consumption of resources and threats to the Tangible and Intangible Heritage are high.

In emerging countries, in fact, even more than in other places, tourism replaces traditional productive activities, modifies the local context, and increases the consumption of already scarce resources, such as water and energy. In addition, it concentrates the economic benefits on a small number of large operators (Bimonte and Punzo 2004).

The resident population has to deal with heavy modifications to the territory, especially in places where mass tourism is more significant. In these contexts, the territory bends to the needs of tourists. In addition, local inhabitants have to face a decrease in their spending power following the increase in costs to adapt to the spending standards of tourists from richer countries.

The UNESCO "brand" may not have positive effects just because the inclusion of a site in the World Heritage List can cause an increase in tourist flows.

Tourism is one of the main drivers for many countries to apply for inclusion in UNESCO's World Heritage list. Inclusion in the World Heritage List brings hitherto little-known sites to the international spotlight, opening them to tourism from "rich" countries (Giambruno et al. 2018).

These processes cause an acceleration in the consumption of Cultural Heritage and increase the cost of living in the WHS areas.

These are the possible effects of tourism development. In this framework, the assumptions and objectives of this work can be better understood. We hope our work can help avoid the mistakes made in those known as First-world countries which are now dealing with the loss of Heritage and the need for sustainable development of the territories.

References

Bimonte S, Punzo LF (2004) A proposito di capacità di carico turistica. Una breve analisi teorica. In: EdATS working papers series, n. 4

European Commission EU, Directorate-General for Education, Youth, Sport and Culture, Sustainable cultural tourism, Publications Office (2019). https://data.europa.eu/doi/10.2766/400886. Accessed 5 July 2022

Giambruno M, Gabaglio R, Pistidda S (2018) Patrimonio culturale e Paesi emergenti. Altralinea Edizioni, Firenze

ICOMOS (1993) Guidelines for education and training in the conservation of monuments, ensembles, and sites. https://www.icomos.org/en/charters-and-texts/179-articles-en-francais/res sources/charters-and-standards/187-guidelines-for-education-and-training-in-the-conservation-of-monuments-ensembles-and-sites. Accessed 15 Jun 2022

ICOMOS (2002) International cultural tourism committee (ICTC), International cultural tourism charter. https://www.icomosictc.org/p/international-cultural-tourism-charter.html. Accessed 21 Jun 2022

Musso SF (2015) "Inheriting" our cultural heritage: changes of paradigm of conservation. Int J Arch Plan 2(Issue 2):84–103

UNEP United Nations Environment Program and World Tourism Organization (2005) Making tourism more sustainable. A guide for policy makers. https://wedocs.unep.org/handle/20.500.11822/8741. Accessed 10 Jun 2022

UNWTO (2013) Sustainable tourism for development guidebook. Enhancing capacities for sustainable tourism for development in developing countries. https://www.e-unwto.org/doi/book/10.18111/9789284415496. Accessed 10 Jun 2022

Printed in the United States
by Baker & Taylor Publisher Services